SCOPE 17

Some Perspectives
of the
Major Biogeochemical Cycles

Executive Committee of SCOPE

President: Professor G. White, Institute of Behavioral Science, University of Colorado, Boulder, Colorado 80309, U.S.A.

Vice-President: Professor S. Krishnaswamy, School of Biological Sciences, Madurai University, Madurai 625021, India.

Vice-President: Professor G. A. Zavarzin, Institute of Microbiology, USSR Academy of Sciences, 117312 Moscow, U.S.S.R.

Secretary-General: Dr. F. Fournier, Inspecteur Général de Recherches, ORSTOM, 24 rue Bayard, 75008 Paris, France.

Treasurer: Dr. A. H. Meyl, Deutsche Forschungsgemeinschaft, Kennedyalle 40, 53 Bonn, West Germany.

Members

Professor M. A. Ayyad, Botany Department, Faculty of Science, University of Alexandria, Moharran Bay, Alexandria, Egypt.

Dr. H. Egan, Laboratory of the Government Chemist, Cornwall House, Stanford Street, London SE1 9NQ, U.K.

Professor V. Landa, Deputy Secretary General, Czechoslovak Academy of Sciences, Národni 3, Prague 1, Czechoslovakia.

Professor J. W. M. La Rivière, International Institute for Hydraulic and Environmental Engineering, Oude Delft 95, P.O. Bx 3015, 2601 DA Delft, The Netherlands.

Professor T. Rosswall, Department of Microbiology, Swedish University of Agricultural Sciences, S-75007 Uppsala, Sweden.

Editor-in-Chief

Professor R. E. Munn, Institute for Environmental Studies, University of Toronto, Toronto, Canada, M5S 1A4.

SCOPE 17

Some Perspectives of the Major Biogeochemical Cycles

Edited by
Gene E. Likens
*Section of Ecology and Systematics,
Division of Biological Sciences,
Cornell University, Ithaca, New York, USA*

Published on behalf of the
Scientific Committee on Problems of the Environment (SCOPE)
of the
International Council of Scientific Unions (ICSU)
by

JOHN WILEY & SONS
Chichester · New York · Brisbane · Toronto

Copyright © 1981 by the
Scientific Committee on Problems of the Environment (SCOPE)

All rights reserved.

No part of this book may be reproduced by any means, nor transmitted, nor translated into a machine language without the written permission of the copyright holder.

British Library Cataloguing in Publication Data:

Some perspectives of the major biogeochemical
   cycles.–(SCOPE; 17)
   1. Biogeochemical cycles
   I. Likens, Gene E.   II. Scientific Committee
on Problems of the Environment
574.5'222   QH344   80-42017
ISBN 0 471 27989 7

Typeset in Great Britain by Activity, Teffont, Salisbury, Wilts and printed by Pitman Press, Bath.

International Council of Scientific Unions (ICSU)
Scientific Committee on Problems of the Environment (SCOPE)

SCOPE is one of a number of committees established by a non-governmental group of scientific organizations, The International Council of Scientific Unions (ICSU). The membership of ICSU includes representatives from 68 National Academies of Science, 18 International Unions, and 12 other bodies called Scientific Associates. To cover multidisciplinary activities which include the interests of several unions, ICSU has established 10 scientific committees, of which SCOPE, founded in 1969, is one. Currently, representatives of 34 member countries and 15 Unions and Scientific Committees participate in the work of SCOPE, which directs particular attention to the needs of developing countries.

The mandate of SCOPE is to assemble, review, and assess the information available on man-made environmental changes and the effects of these changes on man; to assess and evaluate the methodologies of measurement of environmental parameters; to provide an intelligence service on current research; and by the recruitment of the best available scientific information and constructive thinking to establish itself as a corpus of informed advice for the benefit of centres of fundamental research and of organizations and agencies operationally engaged in studies of the environment.

SCOPE is governed by a General Assembly, which meets every three years. Between such meetings its activities are directed by the Executive Committee.

> R. E. Munn
> Editor-in-Chief
> SCOPE Publications

Executive Secretary: Dr V. Smirnyagin

Secretariat: 51 Bld de Montmorency
              75016 PARIS

SCOPE 1: Global Environmental Monitoring 1971, 68pp (out of print)

SCOPE 2: Man-Made Lakes as Modified Ecosystems, 1972, 76pp

SCOPE 3: Global Environmental Monitoring System (GEMS): Action Plan for Phase 1, 1973, 132pp

SCOPE 4: Environmental Sciences in Developing Countries, 1974, 72pp

Environment and Development, proceedings of SCOPE/UNEP Symposium on Environmental Sciences in Developing Countries, Nairobi, February 11-23, 1974, 418pp

SCOPE 5: Environmental Impact Assessment: Principles and Procedures, 1975, 160pp

SCOPE 6: Environmental Pollutants: Selected Analytical Methods, 1975, 277pp (available from Butterworth & Co. (Publishers) Ltd., Sevenoaks, Kent, England)

SCOPE 7: Nitrogen, Phosphorus, and Sulphur: Global Cycles, 1975, 192pp (available from Dr Thomas Rosswall, Swedish Natural Science Research Council, Stockholm, Sweden)

SCOPE 8: Risk Assessment of Environmental Hazard, 1978, 132pp

SCOPE 9: Simulation Modelling of Environmental Problems, 1978, 128pp

SCOPE 10: Environmental Issues, 1977, 242pp

SCOPE 11: Shelter Provision in Developing Countries, 1978, 112pp

SCOPE 12: Principles of Ecotoxicology, 1979, 372pp

SCOPE 13: The Global Carbon Cycle, 1979, 491pp

SCOPE 14: Saharan Dust: Mobilization, Transport, Deposition, 1979, 320pp

SCOPE 15: Environmental Risk Assessment, 1980, 176pp

SCOPE 16: Carbon Cycle Modelling, 1981, 404pp

SCOPE 17: Some Perspectives of the Major Biogeochemical Cycles, 1981, 175pp

Contents

Foreword ix
Preface xi
List of Contributors xiii

SECTION I : BIOGEOCHEMICAL CYCLES OF CARBON, NITROGEN AND SULPHUR

1 Factors controlling global climate of the past and the future 3
  E. T. Degens, H. K. Wong, and S. Kempe

2 The biogeochemical nitrogen cycle 25
  T. Rosswall

3 Current problems related to the atmospheric part of the sulphur cycle ... 51
  H. Rodhe

4 The global biogeochemical sulphur cycle 61
  M. V. Ivanov

SECTION II: INTERACTIONS BETWEEN MAJOR BIOGEOCHEMICAL CYCLES

5 Chemical coupling of the nitrogen, sulphur and carbon cycles
  in the atmosphere 81
  D. H. Ehhalt

6 Interactions between major biogeochemical cycles in terrestrial
  ecosystems 93
  G. E. Likens, F. H. Bormann, and N. M. Johnson

7 Interrelationships between the cycles of elements in freshwater
  ecosystems 113
  D. W. Schindler

8 Interactions between major biogeochemical cycles in marine
  ecosystems 125
  R. Wollast

SECTION III: SOCIO-ECONOMIC IMPACTS ON BIOGEOCHEMICAL CYCLES

9 Socio-economic impacts of the effects of man on biogeochemical
   cycles: sulphur 145
   G. Persson

10 Socio-economic impacts of carbon dioxide induced climatic changes and the
   comparative chances of alternative political responses—prevention, compensation, and adaptation 157
   K. M. Meyer-Abich

Index 171

Foreword

> 'There is no national science just as there is no national multiplication table; what is national is no longer science'.
>
> *A. P. Chekov* (1860-1904)

The Scientific Committee on Problems of the Environment (SCOPE) has since 1974 conducted a project on biogeochemical cycles. The first phase consisted of an assessment of the global cycles of carbon, nitrogen, phosphorus, and sulphur. Based on the findings of this first phase, SCOPE decided to continue its activities in this highly important scientific field. The second phase led to the establishment of the SCOPE/UNEP International Nitrogen Unit at the Royal Swedish Academy of Sciences in Stockholm, a Carbon Unit at the University of Hamburg with auxiliary units in Stockholm, Bruxelles, Essen, and Woods Hole, and a Sulphur Unit at the Institute of Biochemistry and Physiology of Microorganisms of the USSR Academy of Sciences in Pushchino.

The third phase of the SCOPE project on biogeochemical cycles will look into the problems involved in understanding how the various cycles interact.

The 4th General Assembly of SCOPE, held at the Royal Swedish Academy of Sciences in Stockholm in June 1979, gave a valuable mid-second-phase opportunity to address the advancement of knowledge concerning the major biogeochemical cycles. This was considered an opportunity not so much for looking at what had been done but more to look ahead and to address the question of the importance of the biogeochemical cycles also in a socio-economic perspective.

The first session of the scientific meeting on biogeochemical cycles was devoted to interactions between the cycles. There are many such important interactions and the individual elements must not be treated too much as separate entities without due consideration to the close links which were part of the second phase of the SCOPE programme, while the last session concerned itself with socio-economic aspects of the cycles.

At the time of the SCOPE General Assembly in Stockholm, a joint statement was issued by Dr. M. K. Tolba, Executive Director of the United Nations Environment Programme, and Professor G. F. White, President of SCOPE. This statement drew the attention to 'the fundamental scientific importance of understanding the biogeochemical cycles which link and unify the major chemical and biological processes of the earth's surface and the atmosphere'. The statement invited scientists in various disciplines to contribute to the design and execution of a collective endeavour to establish the essential basis for an understanding of the biogeochemical cycles as a global life-support system. The importance of undertaking such a concerted effort should be evident from this volume and from other publications from the SCOPE project on biogeochemical cycles as well as from a rapid increase in other scientific publications on this topic.

Department of Microbiology
Swedish University of Agricultural Sciences,
Uppsala, Sweden

T. ROSSWALL
Swedish SCOPE
Committee,
Project Coordinator,
SCOPE/UNEP International
Nitrogen Unit

Preface

This little book is a first attempt to consider the biogeochemistry of the globe from a holistic point of view. To date investigations have been focused on the flux and reservoir amounts of individual substances, primarily carbon, nitrogen, phosphorus, sulphur, and water because of their major biological importance. The Scientific Committee on Problems of the Environment (SCOPE) has initiated and sponsored numerous workshops and symposia on these individual cycles and several important publications have emerged. However, it has become increasingly apparent that a realistic understanding of individual cycles is not possible in isolation from other interacting cycles, and although integrative analyses are awesome, it is clear that attempts must be made to take a more holistic approach to global biogeochemistry. This new approach is especially important now because of man's increasing alteration of global cycles. Distortion of cycles by human activities may now equal or exceed present-day natural fluxes for certain elements (e.g. sulphur and carbon) or processes (e.g. erosion). Currently two major, widespread environmental problems, carbon dioxide effects on climate and acid precipitation, illustrate the urgency for quantitative answers relative to changes in the atmospheric emissions of carbon, sulphur, and nitrogen due to increased combustion of fossil fuels. These perturbations affect a variety of other cycles and, obviously, other cycles affect them, all combining to affect human welfare.

The papers of this symposium, presented in Stockholm during June of 1979, point to some of the interactions, raise some of the questions and problems of empirical measurement and socio-economic impact, provide some future directions for research and monitoring, and list some goals. It will remain for future research and publications to provide quantitative answers to these questions on a global scale.

February 1980 GENE E. LIKENS

List of Contributors

F. H. Bormann	School of Forestry and Environmental Studies, Yale University, New Haven, Connecticut 06511, USA
E. T. Degens	Geological-Palaeontological Institute, University of Hamburg, Bundesstrasse 55, D-2000 Hamburg 13, FRG
D. H. Ehhalt	Institute of Atmospheric Chemistry, 5170 Jülich, FRG
M. V. Ivanov	Institute of Biochemistry and Physiology of Microorganisms, USSR Academy of Sciences, Pushchino-on-Oka, Moscow Region, USSR
N. M. Johnson	Department of Earth Sciences, Dartmouth College, Hanover, New Hampshire 03755, USA
S. Kempe	Geological-Palaeontological Institute, University of Hamburg, Bundesstrasse 55, D-2000 Hamburg 13, FRG
G. E. Likens	Section of Ecology and Systematics, Division of Biological Sciences, Cornell University, Ithaca, New York 14850, USA
K. M. Meyer-Abich	Arbeitsgruppe Umwelt, Gesellschaft, Energie (AUGE), University of Essen, Universitätsstrasse 12, D-4300 Essen 1, FRG
G. Persson	The National Swedish Environment Protection Board, Fack, S-171 20 Solna, Sweden
H. Rodhe	Department of Meteorology, University of Stockholm, Arrhenius Laboratory, Fack, S-10691 Stockholm, Sweden
T. Rosswall	Department of Microbiology, Swedish University of Agricultural Sciences, S-75007 Uppsala, Sweden
D. W. Schindler	Department of Fisheries and Oceans, Freshwater Institute, University of Manitoba, 501 University Crescent, Winnipeg, Manitoba R3T 2N6, Canada
R. F. Wollast	Laboratory of Industrial Chemistry, University of Brussels, Avenue F-D Roosevelt 50, B-1050 Brussels, Belgium
H. K. Wong	Geological-Palaeontological Institute, University of Hamburg, Bundesstrasse 55, D-2000 Hamburg, FRG

SECTION I

Biogeochemical Cycles of Carbon, Nitrogen, and Sulphur

Some Perspectives of the Major Biogeochemical Cycles
Edited by Gene E. Likens
© 1981 SCOPE

CHAPTER 1

Factors Controlling Global Climate of the Past and the Future

E. T. DEGENS, H. K. WONG, AND S. KEMPE

Geological-Paleontological Institute, University of Hamburg, Federal Republic of Germany

ABSTRACT

Viewpoints expressed in the literature on causes of climatic change are examined and critically assessed. It appears that global climate is a result of a series of factors that operate on different time scales. For this reason we distinguish between climate curves which show changes that have periodicities of millions, thousands, or hundreds of years. All three curves are superimposed and global climate at any one time in the geological past is an expression of this superposition. It is concluded that plate tectonics is the principal long-term regulating factor of global climate by controlling land-sea ratio and albedo. Orbital periodicities of the earth which fluctuate at medium-term time scales (10^4–10^6 years) control the solar radiation curve. The main contributing factor for climatic variation over the past few hundred to few thousand years is short-term periodic changes in the luminosity of the sun and in the volcanic dust concentrations of the atmosphere. In contrast, CO_2 in the atmosphere has not been a climate-controlling device in the past, but will become the principal anthropogenic agent over the next hundred years.

1.1 INTRODUCTION

Our planet Earth has undergone, from the time of its formation, a series of warm periods and ice ages. Since life appeared on land 500 million years ago, however, the global average temperature has fluctuated only by a few degrees Centigrade, although glaciations have occurred during the Permian [270-230 m.y. B.P. (million years before present)] and the Quaternary (2-0 m.y. B.P.). Closer examination of the geological record reveals that the earth was always warm enough to support life, and excursions into ice ages were only of short duration (10^6 yr). Yet, their impact on terrestrial fauna and flora was substantial due to a shift in the positions of climatic zones during the cold interludes.

At present we live in a moderately warm interglacial stadial of an ice age or, expressed differently, in an overall climatic situation which in a geological scale, has little chance of persisting in time. Actually, over the past few hundred thousand years, the climatic pendulum has swung several times almost over the full range from the extreme warm to the extreme cold. These events are registered in sediments in the form of distinct markers and our main objective in this paper will be to use such indicators to find out more about the reasons behind climatic changes throughout the history of the earth.

For the purpose of this discussion we shall distinguish climatic factors which operate on long-term (10^6-10^8 yr), medium-term (10^4-10^6 yr) and short-term (10-10^4 yr) basis. These time scales represent the length of galactic cycles, the period of planetary orbital cycles, and the time span of solar cycles. The climate at any one time is then the superposed effect of all the climatic factors, each operating at its respective time scale.

CO_2 in the atmosphere will be dealt with separately, since it has not been a climate-controlling factor in the past, but may become one over the next few hundred years.

1.2 LONG-TERM FACTORS

1.2.1 Galactic Model

The length of the present galactic year is estimated to be 274 m.y. (Innanen, 1966). This order of magnitude has led to the belief that there is a correlation between the last ice ages in the Quaternary, Permian, and Eocambrian, and the passage of the sun through compression lanes in the spiral arms of the galaxy. Hoyle and Lyttleton (1939) suggested that the additional accretion of interstellar gas would increase the luminosity of the sun which would in turn enhance precipitation and the accumulation of ice on earth. By studying the density of interstellar gas in the vicinity of the solar system, Dennison and Mansfield (1976) could not detect any accumulation of material dense and near enough in space to be responsible for the last glaciation. Steiner (1978) on the other hand presented six lead-isotope events (major geologic disturbances in the uranium/lead radioactive system), two of which he correlated with two dated Precambrian glaciations (940-950 m.y. and 2290 m.y. B.P.). He thereby predicted that there were at least four more Precambrian glaciations, corresponding to the four remaining lead-isotope events. By employing a rather flexible galactic model in which the sun gradually spirals inwards, following an excentric orbit, Steiner was able to place all known (and predicted) glaciations on to one model curve of the same galactic parameter. According to Steiner, glaciation occurs when the solar system reaches its apogalacticum (galactic parameter at a minimum). At the perigalacticum (near the centre of the galaxy), the earth experiences warm periods like those of the Jurassic and the Cretaceous.

1.2.2 Geotectonic Model

A Background

According to plate tectonics, the surface of the earth may be divided into a number of lithospheric plates which move relative to one another. Two plates moving away from each other generate an accreting boundary, along which material of the upper mantle upwells to form a mid-oceanic ridge system. Two converging plates, on the other hand, result in the subduction of one of them in the form of a deep-sea trench, or in the consumption of both of them via mountain-building. Two plates gliding by each other produce a transform fault, where little deformation is to be observed.

These plate motions, in particular place accretion and subduction of an oceanic ridge, have a profound effect on global climate. Firstly, by producing variations in the volume of the mid-oceanic ridge system and hence different rates in the rise and fall of sea-level, they give rise to world-wide transgressions and regressions. In turn, changes result in the ratio of land area to ocean area, the amount of cloud cover, the oceanic circulation pattern, and thereby the global albedo. Secondly, plate motions alter the latitudinal distribution of continents. The percentage of land concentrated within the tropics and in the polar regions therefore changes, causing albedo variations and changes in the position of the large cloud masses. Consequently, perturbations of the albedo and medium-term climatic changes are produced. We shall discuss each of these effects in turn.

B Transgressions, Regressions, and Global Climate

Eustatic sea-level changes have been attributed to various causes. The most effective of these are sudden, catastrophic events such as glaciation and deglaciation and dessication and flooding of small ocean basins. They are, however, ephemeral. Potentially, the fastest way to change sea-level on a long-term basis is to change the volume of the mid-oceanic ridge system (Hallam, 1963; Russel, 1968; Menard, 1969; Valentine and Moores, 1972). Other possibilities involving the production of juvenile water at active ridge areas, the continuing differentiation of the lithosphere (which alters the volume capacity of the ocean basins), variations in sedimentation, and crustal shortening through orogeny produce considerably smaller rates of sea-level change (Hays and Pitman, 1973; Pitman, 1978).

After hot upper mantle material is accreted to a plate, it moves away as part of the plate. In the process of doing so, the material cools and subsides. It is this subsidence that bestows a definite depth-age relationship upon the oceans of the world, regardless of whether the ocean has been generated at a fast spreading ridge (e.g. The Pacific), a slowly spreading ridge (e.g. the Atlantic), a ridge spreading with various rates at various times (e.g. the Indian Ocean), or at a ridge that has ceased spreading a long time ago, e.g. the Labrador Sea (Sclater *et al.*, 1971; Lister, 1972;

Sclater and Dietrick, 1973; Parker and Oldenburg, 1973; Oldenburg, 1975; Tréhu, 1975). By making certain reasonable assumptions about the physical and thermal properties of the lithosphere, it has been shown (Parsons and Sclater, 1977) that to a first approximation, the depth of the sea floor varies as the square root of its age, t, for the age range of 0 to 70 m.y.:

$$d(t) = 2500 + 350(t)^{1/2} \text{ m} \quad (t \text{ in m.y. B.P.}) \tag{1}$$

For an older ocean floor, the depth is given by:

$$d(t) = 6400 - 3200 \exp(-t/62.8) \text{ m} \tag{2}$$

Knowing the sea floor spreading history of the oceans, knowing the ocean depth as a function of age, and knowing the length of actively spreading ridge axes at any point in geological time, we can compute the volumetric change in the mid-oceanic ridge system as a function of time. By assuming that the volume of seawater has remained constant, the computed volumetric change can be interpreted as an inverse change in oceanic basin capacity. Thus, an oceanic ridge volume increase of ΔV would mean a decrease in oceanic basin capacity of ΔV.

To translate volumetric changes in oceanic basin capacity to sea-level fluctuations, two corrections must be made (Hays and Pitman, 1973). The first is the isostatic adjustment of the oceanic basins relative to the continents. When water depth increases by an amount h, the ocean floor subsides a distance d, whereby $h = 3.4 d$, when the upper mantle density is assumed to be 3.4 g cm^{-3}. The change in continental freeboard is hence $h-d = 0.7 h$. The second correction arises because as sea-level rises, more surface area becomes water-covered. About one sixth of the earth's surface (8.5×10^7 km^2) lies between 0 and 500 m (Sverdrup et al., 1942). Assuming that as sea-level rises the additional area flooded by the sea increases linearly (1.7×10^5 km^2 per each metre rise in sea-level), the actual sea-level change $0.7 h$ (in m) can be calculated from

$$\Delta V = hA_0 + 170(0.7 h)^2/2 \tag{3}$$

where ΔV (km^3) is the change in volume of the mid-oceanic ridge system and A_0 (360×10^6 km^2) the present-day area of the oceans.

Using the method outlined above, Pitman (1978) computed the sea-level curve from the upper Cretaceous (85 m.y. B.P.) to mid-Miocene (15 m.y. B.P.) (Figure 1.1). From the mid-Miocene on, glaciation has dominated fluctuations in the sea-level curve, so that the present method is no longer applicable. Two points are of particular importance. Firstly, during the entire period 85–15 m.y. B.P., there was a steady fall in sea-level at a rate equal to or less than 0.7 cm/10^3 yr. Sea-level dropped slowly from 85 to 65 m.y. B.P., quite rapidly through Paleocene and early Eocene, less rapidly during late Eocene, and more rapidly again in the Oligocene. No sea-level rise could be deduced. Secondly, sea-level was about 350 m higher than today in late Cretaceous time. This level is consistent with paleogeographic data (Ronov, 1968) and data on the position of the late Cretaceous shoreline (Sleep, 1976). It

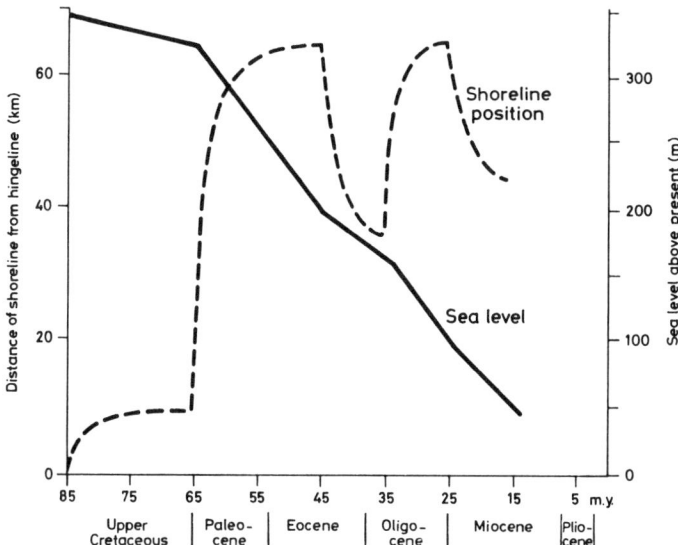

Figure 1.1 Change of sea-level due to volumetric changes in the mid-oceanic ridge system from 85 to 15 m.y. B.P. (million years before present) (solid line). From 15 m.y. B.P. onwards, glacial events determine sea-level curve. Broken line gives the distance of the shoreline from the hinge line (after Pitman, 1978)

also implies that approximately 35 per cent of the present land surface was covered by water. Since water has a much larger thermal capacity compared to rocks on land, this increase in area of sea relative to land must have had a tremendous moderating and stabilizing effect on world climate. In particular, as the extent of epicontinental seas rapidly expanded, convective transfer of heat between low and high latitudes was greatly facilitated, so that a more benign climate was enhanced.

For a long time, it has been believed that transgressions are due to a eustatic sea-level rise and regressions to a eustatic fall. Such a scheme, however, need not necessarily be correct, particularly because a eustatic fall due to glaciation between 85 and 15 m.y. B.P. is not likely. We observe that along Atlantic-type continental margins, kilometres of seaward-thickening, stratified sedimentary sequences overlie a subsided, faulted basement. These sequences are deposited in their entirety within several hundred metres of sea-level. Thus, the sedimentary environment is characterized by rapid subsidence, perhaps at a rate of 2 cm/10^3 yr, so that despite a continuous fall in sea-level, the shoreline remains confined to the shelf. To quantify this interplay of subsidence, sedimentation, and sea-level change, we assume that subsidence occurs about a fixed landward hinge line, and that its rate is constant. Furthermore we assume that sedimentation varies spatially in such a way that the slope of the coastal plain shelf is constant despite subsidence, i.e. sediment infill has kept pace with the subsiding margin. Following Pitman (1978), the rate of change of position of the shoreline (dx/dt) may now be expressed as:

$$S_L \frac{dx}{dt} = R_{SL} - \frac{xR_{ss}}{D} + S \qquad (4)$$

where S_L = slope of the surface of the coastal plain and shelf,
R_{SL} = rate of sea-level change,
x = distance of shoreline from hinge line,
R_{ss} = subsidence rate at shelf edge,
D = distance of shelf edge from hinge line,
and S = additional uniform sedimentation, to allow for possible sedimentation on the coastal plain.

Integrating gives:

$$x = \frac{D}{R_{ss}}(R_{SL} + S) - \left(\frac{R_{SL}D}{R_{ss}} + \frac{SD}{R_{ss}} - x_0\right)\exp\left(-\frac{TR_{ss}}{DS_L}\right) \qquad (5)$$

in which x = position of shoreline after time T and x_0 = position of shoreline at the beginning of interval T.

For large time intervals (10^7 yr), reasonable values of D (250 km), S_L (1/5000), and R_{ss} (2.5 cm/10^3 yr) yield

$$\exp(-TR_{ss}/DS_L) = 1/148$$

so that

$$x \approx \frac{D}{R_{ss}}(R_{SL} + S) \text{ or } R_{SL} \approx \frac{X}{D}R_{ss} - S \qquad (6)$$

i.e. the shoreline tends to stabilize at a point where the rate of change of sea-level equals the difference between the subsidence rate at this point and the sedimentation rate. From the late Mesozoic to the present, excepting glacial effects, the sea-level has been continuously falling. An increase in the rate of fall would require a seaward migration of the shoreline (equation 6) and hence a regression. Likewise, a decrease in sea-level fall would result in a transgression. Hence, transgressions and regressions need not correspond to maxima and minima in sea-level stand. A transgression will occur when sea-level is made to rise more rapidly or to fall more slowly; and if sea-level is made to rise more slowly or fall more rapidly, a regression will result.

In Figure 1.1 the position of the shoreline calculated by equation (5) using the sea-level curve for 85–15 m.y. B. P. is depicted as a broken line. Here the Eocene transgression and the Oligocene regression so often discussed (e.g. Hallam, 1963) are seen to be results of changes in the rate of fall of sea-level.

With the position of the shoreline known, the amount of land surface relative to sea surface can be calculated. Since the reflectivity and its variation with the angle of incident radiation changes from land to ocean, transgressions and regressions cause changes in the global albedo and therefore in the surface temperature of the earth. These changes are, of course, reflected in perturbations in the climate prevailing at the time.

C Changes in the Distribution of Continents and Global Climate

That plate motions alter the spatial distribution of continents and hence the percentage of land within a particular latitudinal zone is obvious (see e.g. Smith et al., 1973; Briden et al., 1974; Smith and Briden, 1977). Here, we shall take a brief look at the effect of these alterations on global climate.

The global albedo A may be expressed as:

$$A = A_C \alpha_C + A_L (1 - \alpha_C) \alpha_L + A_W (1 - \alpha_C)(1 - \alpha_L) \tag{7}$$

where A_C, A_L, and A_W are the average albedos of cloud, land, and water respectively, α_C the fractional cloud cover and α_L the subaerial fraction of the earth's surface. Of the three terms, it appears that the albedo of cloud is the most important (Schneider, 1972). Thus changes in the amount of cloud cover and in its distribution as a result of plate motions must be considered in any theory of climatic change. It has been suggested that the percentage of cloud cover for the earth has remained practically unchanged over evolutionary time periods (10^8-10^9 yr) (Henderson-Sellers, 1979). However, over shorter periods (10^4-10^7 yr), cloud masses could have undergone changes in their distribution pattern with different continental configurations. Areas of high cloud concentrations are not only determined by atmospheric circulation, they are also often found over open oceans. Computer model studies show that a 10 °K increase in surface temperature could lead to a 10 per cent decrease in cloud cover given the present continental configurations, and this decrease would in turn result in a six per cent decrease in the global albedo (Henderson-Sellers, 1979). A 10 °K increase in temperature appears extreme since a 5 °K change is believed typical of glacial-interglacial conditions (Bryson, 1974). However, the interesting point is that such a change produces an inherently stable system, i.e. one with negative feedback in terms of cloud formation. This dynamic stability could be disrupted by possible variations in cloud positions as a result of continental drift. Computations based on different climate models show that such variations tend to perturb the albedo, and though the effects are secondary, they could be important in medium-term global climatic changes.

Changes of continental configuration also effect the land and water terms of equation (7), and these effects are better understood. To a first approximation, we take the land albedo A_L to be constant at 15 per cent, and consider the water albedo A_W to be constant (five per cent) from the equator to 50°N or S, thereafter increasing linearly to 12 per cent at the poles. This change implies that the albedo contrast $(A_L - A_W)$ decreases with increase in latitude.

When plate motions concentrate land masses in the circumpolar regions, the tropics become cloudier and wetter. In the polar regions, replacement of A_W by the slightly higher A_L results in a slight cooling, but the larger albedo contrast within the tropics produces a substantial warming. On the whole, the steeper poleward temperature gradient is overshadowed by a globally warm climate (Cogley, 1979). When continents drift into the tropics, the effect of the albedo contrast produces a rapid cooling at low latitudes and a slight warming near the poles. Global cooling overshadows the reduced poleward temperature gradient.

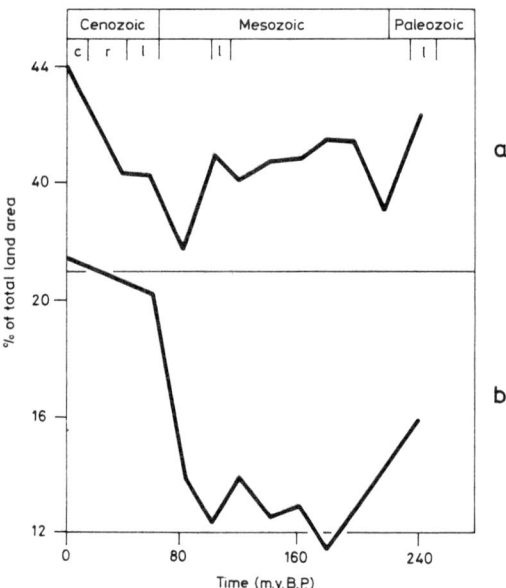

Figure 1.2 (a) Percentage of land area between 30°N and 30°S. (b) Percentage of land mass poleward of 60°N and 60°S; c, r, l stand for continental, regional, and local glaciation, respectively (after Cogley, 1979)

Sellers and Meadows (1975) have shown that land mass movements cause albedo changes, in the sense discussed above, and may be responsible for triggering or reinforcing glacial periods. They state that land mass concentrations near the poles are conducive to glaciations. The argument above (Cogley, 1979) suggests that perhaps concentrations of land both around the poles and in the tropics are necessary, as both represent climatic instabilities.

We plot in Figure 1.2 the percentage of land areas berween 30°N and 30°S (curve a), and polewards of 60°N and 60°S (curve b) respectively. The Cenozoic cooling trend, the glaciations since mid-Miocene, and the end of the extensive Paleozoic glaciations (which is as far back in time as these curves extend), all seem to be accompanied by high percentages of land both in the polar regions and in the tropics.

1.2.3 Effect of a Faster Earth's Rotation Rate in the Geological Past

Growth patterns of shells and corals suggest that in the late Precambrian about 1.5 $\times 10^9$ yr ago, the earth's rotation rate was 2-2.5 times greater than at present (Mohr, 1975). The effect of such a high rotation rate on climate, although speculative, has been evaluated by Hunt (1979) using a numerical model of atmospheric circulation. By assuming that cloud cover, surface albedo, and CO_2 content in the atmosphere were all the same as today, he deduced that a faster rotation of the earth would result in a reduction in the scale size and intensity of the zonal winds, a decrease in poleward heat transport, and a more equatorial location of the tropospheric westerly jet stream. In Precambrian times, therefore, the surface wind stress, being proportional to the square of the wind velocity, would have been drastically reduced. The oceanic gyre would have been weakened and wind-induced vertical mixing in the oceans would have been much less significant, so that the surface isothermal layer would be shallower and warmer. The implication is a diminished poleward transport of heat by the oceans, and hence warmer tropics and a colder polar region. Hunt speculated that for the Precambrian, the climatic changes deduced from a faster earth rotation alone could lead to glaciations. By 6×10^8 yr ago, the much slower rotation rate had improved the climate to such an extent that the maintenance of glacial conditions was no longer possible. The Precambrian glaciations thereby came to an end.

Whether Precambrian glaciations were indeed caused by a faster rotation of the earth may be debatable. The effect of a faster rotation on global climate, however, clearly deserves closer scrutiny.

1.3 MEDIUM-TERM FACTORS

1.3.1 Tectonic Pulses

The long-term geotectonic model was based on constant rates of tectonic subsidence, uplift, and spreading for periods of millions of years. Closer examination reveals that tectonic events occur in pulses which are followed by times of tectonic quiescence. The frequency of such pulses is of the order of ten thousand to a few hundred thousand years. At times they come in series, are globally distributed, and have consequently led to major mountain-building epochs, such as the Caledonian or the Variscian. At other times they are really confined and of shorter duration. Nevertheless, they are accompanied by the same phenomena as discussed earlier, i.e. subsidence, uplift, transgression, or regression, and their impact on regional climate can be substantial. The likelihood even exists that regional tectonics proceeding in 'climate-strategic' parts of the earth may influence global climate through a series of feedback systems. As an example we consider Southeast Europe during the Quaternary.

The Quaternary started with the Akčagylian marine transgression which transformed wide areas of southern Russia into a shallow sea by uniting the Black Sea, the Caspian Sea, and Lake Aral. The sea measured more than 2000 km in a north-south direction and almost 1000 km at its widest east-west opening (Figure 1.3). Shallow water conditions prevailed for almost two million years and the Danube drained into the Caspian Sea. About 400 000 years ago and continuing towards the

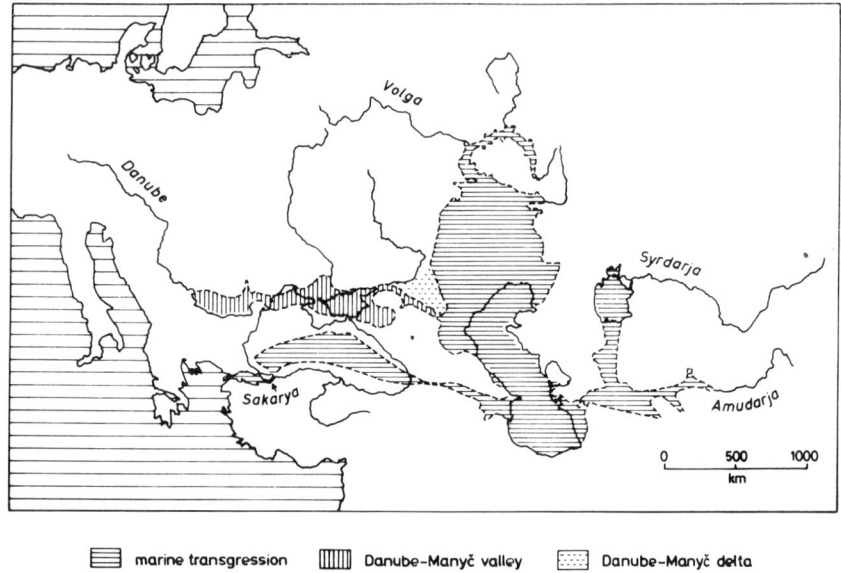

Figure 1.3 Akčagylian transgression in the Black Sea–Caspian Sea–Lake Aral region (after Degens and Paluska, 1979)

present, pulses of tectonic activity caused rapid subsidence of a basin chain extending from the Caspian Sea to the lowlands of the Po valley; rates of subsidence were as high as 1–2 cm yr^{-1} with a maximum about 200 000 years ago (Degens and Paluska, 1979; Paluska and Degens, 1979). Synchronous with the subsidence in the south was a general uplift in parts of Russia, Anatolia, Central Europe, and Scandinavia. For example, within a hundred thousand years Scandinavia became uplifted by one kilometre and more, while at the same time the Caspian and Black Sea basin floors sank by the same amount. This vertical motion provided the kinetic energy for massive erosion and formation of moraines in the aftermath of glacial events. The fact that the past three major ice ages coincide with a major tectonic pulse in Europe suggests that rapid tectonism leading to regional changes in land/water ratio, albedo, topography, orography, and bathymetry may contribute to global climatic alterations due to the 'weather-strategic' position of this area.

1.3.2 Orbital Periodicities of the Earth

From the periodicities in the tilt angle of the rotation axis (41 000 years), the precession of the earth's axis (21 000 years), and the eccentricity of its orbit (93 000 years) Milankovitch deduced a curve spanning the last 300 000 years for the radiation values averaged over the months March to September. Today the solar radiation at 50°N is 847 langley day^{-1} (or 35 447 kJ m^{-2} day^{-1}). For the past 300 000 years, the Milankovitch curve shows fluctuations between −35 and 50 ly day^{-1} with respect to the present day value. These differences in radiation are, according to the Weertman model (1976), sufficient to cause glaciation on northern high latitude continents. The Milankovitch curve suggests that the earth has recently finished a phase for which the solar radiation incident on its surface was very high. At present, this incident radiation is 30 ly day^{-1} less, so that cooling is expected. Weertman's model predicts the beginning of a new ice age within the next several hundred or several thousand years. It should last approximately 60 000 years, i.e. its duration should be similar to the Wisconsin Ice Age. However, Weertman also pointed out that his model requires much higher precipitation rates than those prevalent in the continental polar areas today. Furthermore, he also noted that the occurrence of ice ages in model calculations is largely a consequence of choosing the 'right' model parameters.

Orbital periodicities have recently been suggested by several scientific schools to be the principal cause for climatic change of the type recorded by the glacial-interglacial pattern of the Quaternary. For a comprehensive account see Imbrie and Imbrie (1979).

1.4 SHORT-TERM FACTORS

1.4.1 Volcanic Dust

Large volcanic eruptions eject immense amounts of volcanic dust into the stratosphere. These dust particles increase the reflected portion of the incoming solar radiation, thus effectively diminishing the proportion of solar energy reaching the earth's surface. Consequently, temperatures at ground level become lower, and colder weather is to be expected.

In 1815 Mount Tambora in Sumbawa in Indonesia exploded, ejecting over 100 km^3 of rocks and ash. Large amounts of ash were blasted into the stratosphere, whence it started to distribute around the globe. Throughout May, 1816, the weather in eastern North America and parts of Europe stayed cool to become what was the coldest summer on the temperature record for New Haven (Connecticut) kept at Yale College (Stommel and Stommel, 1979; Figure 1.4). Although whether this apparent cooling can be directly attributed to the Tambora eruption is disputable (Self and Rampino, 1979; Stommel, 1979), that the eruption did play a role in the weather change is perhaps not controversial.

A global cooling was evident in the temperature trend following the eruption of Krakatoa in 1883 (Mitchell, 1977). Likewise, the injection of aerosol into the atmosphere accompanying the burst of Mount Agung on Bali in March, 1963, probably caused low Pacific and high tropical stratospheric temperatures (Newell and Weare, 1976).

Schneider and Mass (1975) calculated temperature changes from a climatic model in which dust concentrations and solar activity were taken into consideration. Solar transmission measurements at the Mauna Loa Observatory carried out after the

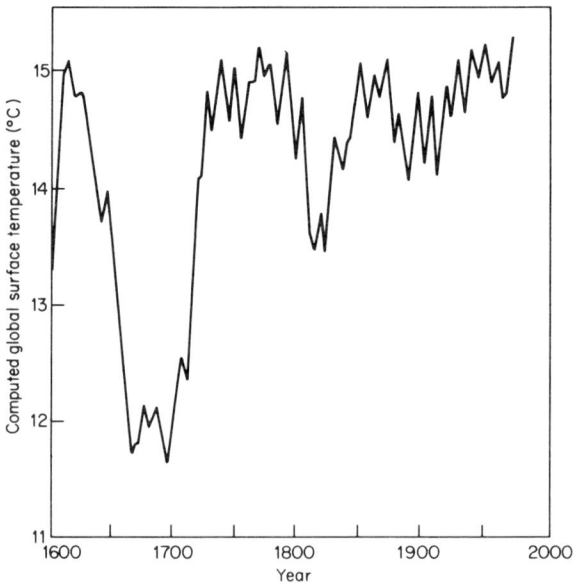

Figure 1.4 Mean June temperature records for New Haven, Connecticut, over a period of 70 years (after Stommel and Stommel, 1979)

Agung eruption showed a reduction in the solar energy directly incident on the earth by two per cent due to volcanic dust. By arguing that nevertheless, three quarters of this missing two per cent eventually reached the earth's surface via scattering, they calculated that the total solar parameter S is reduced by 0.5 per cent as a result of an Agung-scale eruption.

One input function to the climatic model of Schneider and Mass (1975) is the dust veil index obtained empirically from historic records of volcanic eruptions (Lamb, 1970). A second input function is that component of the solar parameter which is dependent on the sunspot number N:

$$S(N) = 1.903 + 0.011\ N^{0.5} - 0.0006\ N \text{ cal cm}^{-2} \text{ min}^{-1}$$

This functional dependence is based on solar constant measurements of Kondratyev and Nikolsky (1970), and its validity is not universally accepted. For the quiet ($N \approx 0$) and the very active sun ($N \approx 200$), it yields values for the sunspot component of the solar parameter, $S(N)$, about two per cent lower than those for a normal solar activity ($N \approx 80$).

Schneider and Mass expressed the solar parameter S as the sum of two factors:

$$S = S(N) + S(D)$$

where $S(D)$ is that component of the solar parameter which is dependent upon dust concentration in the atmosphere. This sum S constitutes the forcing input to their

Factors Controlling Global Climate of the Past and the Future

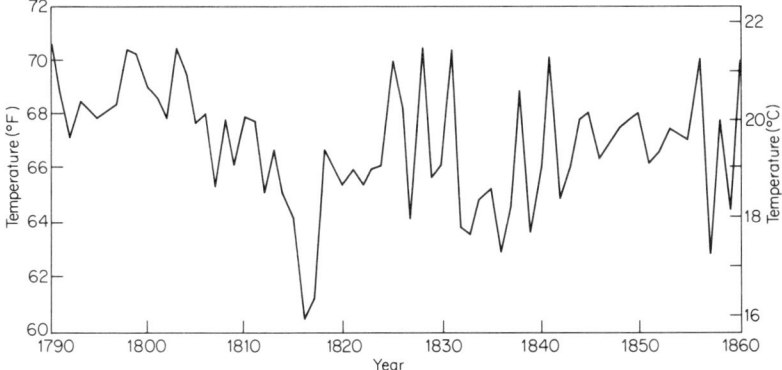

Figure 1.5 Mean global surface temperatures computed by the climate model of Schneider and Mass (1975)

climatic model from which the global surface temperature variation since 1600 can be calculated (Figure 1.5).

The resulting temperature curve is in fair agreement with well-known climatic features, such as the little ice age which is explained by the Mauder minimum in the solar activity, and the cold early eighteen hundred which are explained by both a high dust index and a low solar activity.

$\delta\ ^{18}O$ values (believed to be a measure of temperature) as a function of age have been determined using the Camp Century ice core from Greenland (Dansgaard et al., 1971). This function has been shown to correlate significantly with the envelope of the lunar tidal stress curve at 60°N (Roosen et al., 1976). Roosen et al. (1976) and Lamb suggested that this correlation implies a dependence of climate on lunar tides. Indeed, they argued that volcanic eruptions are statistically apt to occur when high tidal stresses prevail in the lithospheric plates. Such eruptions increase the stratospheric dust content, thereby leading to a cooling in the climatic trend. Coincidence of the 179-year lunar tidal period with the 180-year period of the $\delta\ ^{18}O$ maxima is strong evidence for such a dependence. However, it should be noted that vulcanism (and dust production) is not the only mechanism by which lunar tides might induce periodic climatic variations.

1.4.2 Tidal Forces and Planetary Alignments

Gribbin (1973) observed that the alignment of the outer planets also follows a periodicity of 179 years. He thereby speculated that the tidal forces due to these planets modulate the amplitudes of the individual 11-year sunspot cycles and that the periodicity of this modulation is 180 years. The nature and magnitude of this tidal influence on the sun have as yet not been discussed in a climatological context.

1.4.3 Solar Activity

The only record of solar activity which can be analysed for periodicities is that of sunspots (Schove, 1955). Fourier analysis of this record (Ekdahl and Keeling, 1973) reveals that well-known 11-year sunspot cycle, a cycle of about 80 years, and evidence for the 180-year cycle. Longer cycles can only be deduced from earthbound climatic records with the assumption that the changes found have in fact been produced by the sun. It has often been pointed out (e.g. Smith and Gottlieb, 1975) that the sunspots themselves cannot cause any significant change in the radiation balance of the sun. Even at high sunspot numbers only a 0.1 per cent change in the visible solar flux could result from direct shading by sunspots. The mechanism of climate influence must therefore be more complicated. Dicke (1979) suggested that sunspots are only the surface expressions of a 'solar chronometer' which modulates the luminosity of the sun. This magneto-fluid dynamic oscillator, deeply buried inside the sun, causes magnetic fields to float to the surface of the sun to create sunspots years after the luminosity change occurs. A full magnetic cycle of the sun contains two sunspot cycles, i.e. the magnetic cycle has a length of 22 years. From one 11-year to the next 11-year cycle, the magnetic sign of the sunspots changes: after a cycle with positively charged sunspots on the northern hemisphere of the sun, a cycle with negative spots follows. Epstein and Yapp (1976) found that the deuterium/tritium (D/T) isotope record of a particular bristle cone pine tree shows a 22-year periodicity. By comparing this D/T isotope curve with the sunspot cycle, Dicke (1979) showed that the curves for the latter lagged the former by about 13 years. This time lag may be identical with the time lag between the actual luminosity change of the sun and the appearance of sunspots. A climatic change on the earth reflects of course changes in the solar luminosity.

The evidence for an 11-year frequency in climatic data, however, is intriguing. Drought data have been found to correlate with sunspots (Roberts, 1975) as well as glacier recession, temperatures, and a wealth of other meteorological parameters (Bandeen and Maran, 1975). For Lake Van in eastern Turkey, lake level oscillations have been correlated with sunspot activity (Kempe, 1977). These oscillations not only show a time lag of one year (or following Dicke, 12 years) but evidence for a 10-year solar (?) cycle could be found in the early Holocene varved lake sediments.

Since concentration and generation of ^{14}C in the atmosphere is a function of the intensity of cosmic rays, it might be correct to call ^{14}C periodicities solar. Suess (1970) found periods of 2400 and 405 years. Dansgaard and co-workers (1971) originally found 2100- and 350-year cycles in the oxygen isotope data in the Camp Century ice core, which were standardized to give a time scale with the same periodicities as those of the ^{14}C data. A review of a variety of other cycles of climatic indicators is given by Schove (1978), while Flohn (1978) gives details on acyclic, abrupt events for comparison.

1.5 CO_2 AND CLIMATE

1.5.1 Introductory Remarks

The gigantic climatic experiment mankind is currently undertaking, i.e. the combustion of fossil fuels (which have been accumulated over much of the earth's history) within geologic-zero-time, has focussed our attention on this climatic thermostat. The temperature of the atmosphere is largely determined by the concentration of CO_2, because molecular CO_2 absorbs short wave radiation and releases it as infra-red. Earth is habitable only because of the 300 p.p.m. CO_2 in its atmosphere. If this concentration were substantially lower, then life would have been impossible on this planet because it would have been too cold.

Over geologic history an ingenious sytem has developed whereby CO_2 is redistributed between atmosphere, vegetation, oceans, and rocks by the carbon cycle (Bolin et al., 1979). This cycle works fast for the atmosphere. If no reflux occurred the air would lose its CO_2 to vegetation and ocean within seven years; small surpluses are almost immediately stored away in the oceans. At present perhaps as much as 10 billion tons* of anthropogenic carbon are added annually to the atmosphere in the form of CO_2, five billion tons of it from industrial activities and the rest of it from vegetation and soil destruction (Bolin et al., 1979). About 2.5 billion tons of carbon stay in the atmosphere, corresponding to an increase of about 1 p.p.m. per year (Keeling and Bacastow, 1977; Lowe et al., 1979). The remainder is accounted for by the 'CO_2 buffer capacity' of the earth by way of the carbon cycle. The future atmospheric CO_2 trend is difficult to predict because of uncertainties in the pattern of industrial and agricultural CO_2 production.

1.5.2 Natural Processes

Carbon is stored in the earth's crust principally in the form of coal, oil, natural gas, kerogen (finely disseminated organic matter in sediments), and carbonates. Part of this carbon becomes reactivated by magmatic and volcanic processes and re-enters the atmosphere and hydrosphere mainly as dissolved or gaseous CO_2. Furthermore, erosion exposes fresh rock formations to the action of weathering and liberates additional fossil carbon. In the process of denudation, part of the vegetation cover and soils is removed and carried to the sea as particulate organic matter or in the dissolved form. A substantial fraction of the plant material, however, becomes oxidized to CO_2 and escapes into the air. In the following, we shall briefly examine to what extent natural processes can alter the atmospheric CO_2 level.

Volcanic activity presently produces some 0.05 billion tons of carbon per year, a value over one hundred times less than the estimated annual discharge rate of

---

*Here one billion $\equiv 10^9$

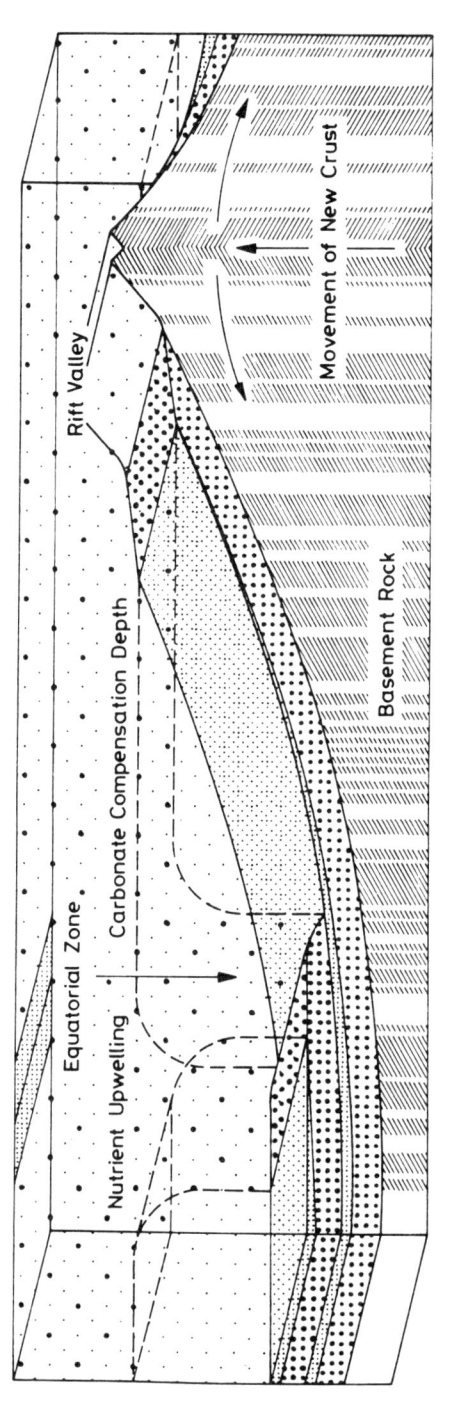

Figure 1.6 Dynamic model (schematic) showing the effect of the carbonate compensation depth (CCD) on the kinds of bottom sediments that accumulate (after Heezen and Macgregor, 1973). Note the movement of new crust which is indicated by a vertical striation pattern reflecting magnetic reversals. Sea floor spreading will move areas into and out of the influence of the CCD. If euxinic conditions develop, due to density stratification, the lysocline and CCD will gradually move up towards the pycnocline; under such conditions a hiatus can be created in abyssal regions that are far removed from land (Degens and Stoffers, 1976)

man-made CO_2. Tectonism and volcanic activities were episodic in the geological past, being linked to the rate of plate accretion and subduction. We live today in a tectonically active phase which could imply that the above volcanic production figure of 0.05 billion tons carbon is a high one on the scale of CO_2 emissions. An additional factor is worth mentioning. Magmatic CO_2 released from the ocean floor into the deep sea can be incorporated into the hydrosphere without substantial changes in the CO_2 partial pressure of the atmosphere. The carbonate compensation depth in the ocean simply adjusts itself in accordance with this development (Figure 1.6).

When the sea floor lies at a level beneath the lysocline, below which carbonates do not exist in the solid phase, carbonaceous sediments cannot be deposited. The depth of the lysocline therefore determines the amount of carbonate which can be removed from the water column. On the other hand, interactions at the air-sea interface control the release of CO_2 into the atmosphere and the biosphere. Things appear quite harmless at first sight. Changes in water density induced by temperature or salt content stratify the sea in mid-water. However, if this stratification is maintained for an extended period of time, mixing of water masses is restricted. The stratification boundary may become so stabilized that it only moves up and down in response to changes in water temperature, tectonic activities, or other environmental disturbances; but it may rarely break up entirely (Degens and Stoffers, 1976, 1977). In such a situation, molecular oxygen would remain abundant in the upper layer but would gradually drop to zero below the density boundary. In short, a euxinic environment (oxygen being absent) is created.

The oscillating pycnocline (stratification boundary due to a rapid density increase with depth) constitutes one of the most powerful means of keeping a status quo for the CO_2 system. We illustrate schematically in Figure 1.7 the operational principle, and its effect on carbonate and sapropel formation.

Oscillating lysoclines and pycnoclines maintain a kind of steady state and shield the CO_2 system from perturbations coming from the atmosphere (air temperature), the lithosphere (volcanic degassing), and the biosphere (primary productivity and decomposition). ^{13}C measurements on tree rings (Freyer, 1978; Stuiver, 1978) suggest that carbon^{-13} isotopic composition of atmospheric CO_2 has decreased by about 2 per mil (parts per thousand) over the past hundred years due to the alteration of the natural CO_2 isotopic composition by the addition of fossil fuel CO_2 (Suess, 1955). The pre-industrial atmosphere has a δ ^{13}C of -5 per mil which conforms with the δ ^{13}C of CO_2 presently evolving from the mantle (Taylor et al., 1967). These data suggest that the 'master switch' for CO_2 control lies deep inside the earth. In view of the huge carbon reservoir in the crust and mantle (Bolin et al., 1979), the dynamics of crustal movement (Engel et al., 1974), and the buffer capacity of the ocean (Takahashi, 1975; Broecker and Takahashi, 1978), this interpretation appears to be realistic.

Thanks to the ocean, carbon dioxide levels in the atmosphere have probably remained fairly uniform, at least since life appeared on earth. Consequently, CO_2

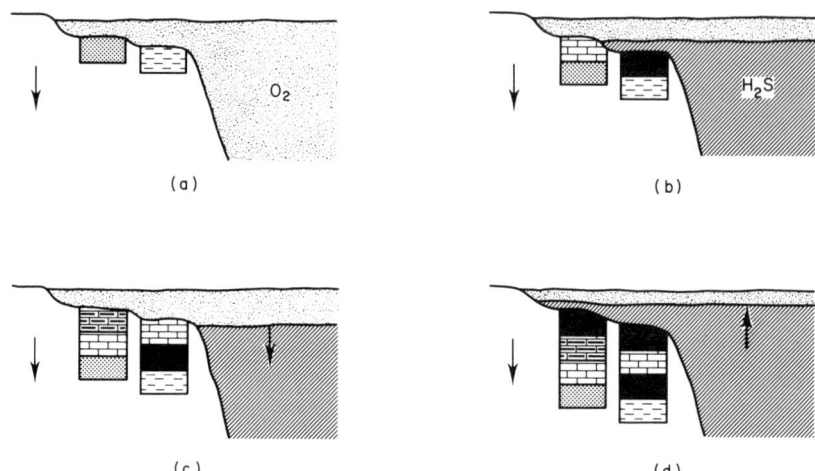

Figure 1.7 Formation and evolution of stratified waters (after Degens and Stoffers, 1976): (a) fully oxygenated sea and deposition of sand (dotted) and clay (dashed areas); (b) stratified sea (density boundary in mid-water); carbonates (brick-wall pattern) form above pycnocline and euxinic sediments (black) below pycnocline in anoxic water; (c) lowering of pycnocline will lead to spasmodic deposition of marl (brick-wall dashed pattern) and carbonates; (d) upward progression of pycnocline will extend euxinic conditions to shallower parts of the sea

is believed to have had only a slight impact on paleoclimate (Degens, in press). In contrast, the present man-made input of CO_2 into our atmosphere is so dramatic that serious consequences for climate and environment are to be expected in the near future.

1.6 SYNOPSIS

In the geologic past likely causes of climatic changes were:

(A) Extraterrestrial

1. variations in the properties of interstellar space or gravity due to galactic rotation (length of the galactic year, ca. 270×10^6 years)
2. variation in the rotational rate of the earth in the geologic past
3. periodic or aperiodic changes in the luminosity of the sun (various solar cycles of 11-, 22-, 80-, 180-year periods)
4. periodicities in the rotation of the earth around the sun (Milankovitch curve; 10^4-10^5 year periods)
5. planetary and lunar tidal effects (180-year period).

(B) Terrestrial (Modulating the Incoming Solar Energy)
1. global changes in the ocean to continent ration (10^6-10^7 year range)
2. redistribution of continents into different latitudes and elevations (10^6-10^7 year range)
3. changes in land/water ratio and elevation due to regional tectonic pulses (10^4-10^5 year range) in climate-strategic parts of the earth
4. episodic changes in concentration of volcanic dust in the atmosphere (1-10 year range).

Extraterrestrial causes are generally of a periodic nature, whereas terrestrial ones proceed within a certain time range.

Geological data indicate that the principal 'dictator' of terrestrial climate is geotectonics. Long-term changes can be attributed to global tectonic activities, midterm changes to regional tectonic events, and short-term changes to volcanic dust injections into the upper atmosphere. Superimposed are modulating effects of extraterrestrial origin.

Our planet is habitable only because of the presence of carbon dioxide and water vapour in the atmosphere. Should these substances somehow be removed, the effective surface temperature would be reduced by about $25°C$ (Rasool and de Bergh, 1970). CO_2 must therefore have played a very important role in the stabilization of global temperatures. Over at least the past 500 million years, CO_2 levels in the atmosphere have stayed uniform due to the buffering effect of the carbonate system in the ocean. This situation may, however, change significantly in the near future. While natural processes will continue to influence climate in the ways described earlier, the role of man-made CO_2 is expected to increase, so that it may become the dominant factor of global climate in the next few hundred years.

1.7 ACKNOWLEDGEMENT

We gratefully acknowledge the financial assistance of the Umweltbundesamt, Berlin, and the Deutsche Forschungsgemeinschaft, Bonn-Bad Godesberg.

1.8 REFERENCES

Bandeen, W. R., and Maran, St. P. (eds) (1975) *Possible Relationships between Solar Activity and Meteorological Phenomena*, A symposium held at Goddard Space Flight Center, Greenbelt, Maryland, Nov. 7-8, 1973; 1-263, NASA, Washington.

Bolin, B., Degens, E. T., Kempe, S., and Ketner, P. (eds) (1979) *The Global Carbon Cycle*, SCOPE Report 13, Chichester, New York, Brisbane, Toronto, Wiley, 1-491.

Briden, J. C., Drewry, G. E., and Smith, A. G. (1974) Phanerozoic equal-area world maps, *J. Geol.*, 82, 555-574.

Broecker, W. S., and Takahashi, T. (1978) The relationship between lysocline depth and in situ carbonate ion concentration, *Deep-Sea Res.*, 25, 65-95.

Bryson, R. A. (1975) A perspective on climatic change, *Science*, 184, 753-780.

Cogley, J. G. (1979) Albedo contrast and glaciation due to continental drift, *Nature*, 279, 712-713.
Dansgaard, W., Johnson, S. J., Clausen, H. B., and Langway, Jr., C. C. (1971) Climatic record revealed by the Camp Century ice core, in Turekian, K. K. (ed.) *The Late Cenozoic Glacial Ages*, Yale University, 37-56.
Degens, E. T. (1979) The past of CO_2. *Environm. Intern.* (in press).
Degens, E. T., and Stoffers, P. (1976) Stratified waters as a key to the past, *Nature*, 263, 22-27.
Degens, E. T., and Stoffers, P. (1977) Phase boundaries as an instrument for metal concentration in geological systems, in Klemm, D. D., and Schneider, H. -J. (eds) *Time- and Strata-bound Ore Deposits*, Berlin, Heidelberg, Springer-Verlag, 25-45.
Degens, E. T., and Paluska, A. (1979) Tectonic and climatic pulses recorded in Quaternary sediments of the Caspian-Black Sea region, *Sed. Geol.*, 23, 149-163.
Dennison, B., and Mansfield, V. N. (1976) Glaciations and dense interstellar clouds, *Nature*, 261, 32-34.
Dicke, R. H. (1979) Solar luminosity and the sunspot cycle, *Nature*, 280, 24-27.
Ekdahl, C. A., and Keeling, C. D. (1973) Atmospheric carbon dioxide and radiocarbon in the natural carbon cycle: I. Quantitative deductions from records at Mauna Loa Observatory and at the South Pole, in Woodwell, G. M., and Pecan, E. V. (eds) *Carbon and the Biosphere*, AEC Symposium Series, 30, 51-85, Springfield, Virginia, NTIS US Department of Commerce.
Engel, A. E. J., Itson, S. P., Engel, C. G., Stickney, D. M., and Cray, Jr., E. J. (1974) Crustal evolution and global tectonics: a petrogenic view, *Geol. Soc. Amer. Bull.*, 85, 843-858.
Epstein, S., and Yapp, C. J. (1976) Climatic implications of the D/H ratio of hydrogen in C-H groups in tree cellulose, *Earth Planet. Sci. Lett.*, 30, 252-261.
Flohn, H. (1978) Abrupt events in climatic history, in Pittock, A. B., Frakes, L. A., and Zillman, J. (eds) *Climatic Change and Variability—A Southern Perspective*, Cambridge, Cambridge University Press, 124-134.
Freyer, H. D. (1978) Preliminary evaluation of past CO_2 increase as derived from ^{13}C measurements in tree rings, in Williams, J. (ed.) *Carbon Dioxide, Climate and Society*, Oxford, New York, Toronto, Sydney, Paris, Frankfurt, Pergamon, 69-77.
Gribbin, J. (1973) Planetary alignments, solar activity and climatic change, *Nature*, 246, 453-454.
Hallam, A. (1963) Major epeirogenic and eustatic changes since the Cretaceous, and their possible relationship to crustal structure, *Amer. J. Science*, 261, 397-423.
Hays, J. D., and Pitman, W. C., III (1973) Lithospheric plate motion, sea level changes and their climatic and ecological consequences, *Nature*, 246, 18-22.
Heezen, B. C., and Macgregor, I. D. (1973) The evolution of the Pacific, *Scientific American*, 229(11), 102-112.
Henderson-Sellers, A. (1979) Clouds and the long-term stability of the earth's atmosphere and climate, *Nature*, 279, 786-788.
Hoyle, F., and Lyttleton, R. A. (1939) The effect of interstellar matter on climatic variation, *Proc. Camb. Phil. Soc.*, 35, 405-415.
Hunt, B. G. (1979) The effects of past variations of the earth's rotation rate on climate, *Nature*, 281(5728), 188-191.
Imbrie, J., and Imbrie, K. P. (1979) *Ice Ages: Solving the Mystery*, published by Enslow, Short Hills, New Jersey, and by Macmillan, London, 1-224.
Innanen, K. A. (1966) The sun's orbit in a mass model of the galactic system, *Z. Astrophys.*, 64, 457.

Keeling, C. W., and Bacastow, R. B. (1977) Impact of industrial gases on climate, in *Energy and Climate, Stud. Geophys.*, Washington, D.C., National Academy of Sciences, 72-95.

Kelly, P. M., and Lamb, H. H. (1976) Prediction of volcanic activity and climate, *Nature*, **262**, 5.

Kempe, S. (1977) Hydrographie, Warven-Chronologie und organische Geochemie des Van-Sees, Ost-Türkei, *Mitt. Geol. Paläont. Inst Univ. Hamburg*, **47**, 125-228.

Kondratyev, K. Ya., and Nikolsky, G. A. (1970) Solar radiation and solar activity, *Quart. J.R. Met. Soc.*, **96**, 509-522.

Lamb, H. H. (1970) Volcanic dust in the atmosphere; with a chronology and assessment of its meteorological significance, *Phil. Trans. Roy. Soc. London, A*, **266**, 425-533.

Lister, C. R. B. (1972) On the thermal balance of a mid-ocean ridge, *Geophys. J. Roy. Astron. Soc.*, **26**, 515-535.

Lowe, D. C., Guenther, P. R., and Keeling, C. D. (1979) The concentration of atmospheric carbon dioxide at Baring Head, New Zealand, *Tellus*, **31**, 58-67.

Menard, H. W. (1969) Elevation and subsidence of oceanic crust, *Earth Planet. Sci. Lett.*, **6**, 275-284.

Mitchell, J. M. (1977) The changing climate, in *Energy and Climate, Stud. Geophys.*, Washington D.C., National Academy of Sciences, 51-58.

Mohr, R. E. (1975) Measured periodicities of the Biwabik (Precambrian) stromatolites and their geophysical significance, in Rosenberg, G. D., and Runcorn, S. K. (eds) *Growth Rhythms and the History of the Earth's Rotation*, London, Wiley, 43-56.

Newell, R. E., and Weare, B. C. (1976) Ocean temperatures and large scale atmospheric variations, *Nature*, **262**, 40-41.

Oldenburg, D. W. (1975) A physical model for the creation of the lithosphere, *Geophys. J. Roy. Astron. Soc.*, **43**, 425-451.

Paluska, A., and Degens, E. T. (1979) Climatic and tectonic events controlling the Quaternary in the Black Sea region, *Geol. Resch.*, **68**, 284-301.

Parker, R. L., and Oldenburg, D. W. (1973) Thermal model of ocean ridges, *Nature*, **242**, 137-139.

Parsons, B., and Sclater, J. G. (1977) An analysis of the variation of ocean floor bathymetry and heat flow with age, *J. Geophys. Res.*, **82**(5), 803-827.

Pitman, W. C., III (1978) Relationship between eustacy and stratigraphic sequences of passive margins, *Geol. Soc. Amer. Bull.*, **89**, 1389-1403.

Rasool, S. I., and de Bergh, C. (1970) The runaway greenhouse and the accumulation of CO_2 in the Venus Atmosphere, *Nature*, **226**, 1037-1039.

Roberts, W. D. (1975) Relationships between solar activity and climate change, in Bandeen, W. R., and Maran, St. P. (eds) *Possible Relationships between Solar Activity and Meteorological Phenomena*, Washington, D.C., NASA, 13-24.

Ronov, A. B. (1968) Probable change in composition of seawater during the course of geological time, *Sedimentology*, **10**, 25-43.

Roosen, R. J., Harrington, R. S., Giles, J., and Browning, I. (1976) Earth tides, volcanoes and climatic changes, *Nature*, **261**, 680-682.

Russel, K. L. (1968) Ocean ridges and eustatic changes in sea level, *Nature*, **218**, 861-862.

Schneider, S. H. (1972) Cloudiness as a global climatic feedback mechanism: The effects on the radiation balance and surface temperature of variations in cloudiness, *J. Atm. Sci.*, **29**, 1413-1422.

Schneider, S. H., and Mass, C. (1975) Volcanic dust, sunspots and temperature trends, *Science*, **190**, 741-746.

Schove, D. J. (1955) The sunspot cycle 649 B.C. to A.D. 2000, *J. Geophys. Res.*, **60**, 127-146.

Schove, D. J. (1978) Tree-ring and varve scales combined, c. 13500 B.C. to A.D. 1977, *Paleogeography-Paleoclimatology-Paleoecology*, **25**, 209-233.

Sclater, J. G., and Dietrick, R. (1973) Elevation of mid-ocean ridges and the basement age of JOIDES deep sea drilling sites, *Geol. Soc. Amer. Bull.*, **84**, 1547-1554.

Sclater, J. G., Anderson, R. N., and Bell, M. L. (1971) Elevation of ridges and evolution of the central eastern Pacific, *J. Geophys. Res.*, **76**, 7888-7915.

Self, S., and Rampino, M. R. (1979) Discussion on 'The year without a summer' by H. Stommel and E. Stommel, *Scientific American*, **241**(4), 10.

Sellers, A., and Meadows, A. J. (1975) Long-term variations in the albedo and surface temperature of the earth, *Nature*, **254**, 44.

Sleep, N. H. (1976) Platform subsidence mechanisms and eustatic sea level change, *Tectonophys.*, **36**, 45-56.

Smith, A. G., and Briden, J. C. (1977) *Mesozoic and Cenozoic Paleocontinental Maps*, Cambridge, Cambridge University Press, 1-63.

Smith, A. G., Briden, J. C., and Drewry, G. E. (1973) Phanerozoic world maps, in Hughes, N. F. (ed.) Organisms and continents through time, *Palaeontology Special Papers*, **12**, 1-42.

Smith, E. P., and Gottlieb, D. M. (1975) Solar flux and its variation, in Bandeen, W. R. and Maran, St. P. (eds) *Possible Relationships between Solar Activity and Meteorological Phenomena*, Washington D.C., NASA, 97-117.

Steiner, J. (1978) Lead isotope events of the Canadian shield, ad hoc solar galactic orbits and glaciations, *Precambrian Res.*, **6**, 269-274.

Stommel, H. (1979) Reply to the discussion by S. Self and M. R. Rampino, *Scientific American*, **241**(4), 10.

Stommel, H., and Stommel, E. (1979) The year without a summer, *Scientific American*, **240**(6), 134-140.

Stuiver, M. (1978) Atmospheric carbon dioxide and carbon reservoir changes, *Science*, **199**, 253-258.

Suess, H. E. (1955) Radiocarbon concentration in modern wood, *Science*, **122**, 415-417.

Suess, H. E. (1970) The three causes of the secular carbon-14 fluctuations, their amplitudes and time constants, in Olsson, I. U. (ed.) *Radiocarbon Variations and Absolute Chronology, Twelfth Nobel Symposium*, Uppsala, 1969, published by Almquist and Wiksel, Stockholm, and Wiley, New York.

Sverdrup, H. U., Johnson, M. W., and Fleming, R. H. (1942) *The Oceans*, Engelwood Cliffs, N.J., Prentice-Hall, 1-1087.

Takahashi, T. (1975) Carbonate chemistry of seawater and the calcite compensation depths in the ocean, in Sliter, W. V., Béand, A. W., and Berger, W. H. (eds) Dissolution of deep-sea carbonates, *Cushman Found. Foraminif. Res., Spec. Publ.*, **13**, 11-26.

Taylor, Jr., H. P., Frechen, J., and Degens, E. T. (1967) Oxygen and carbon isotope studies of carbonatites from the Laacher See district, West Germany, and the Alno district, Sweden, *Geochim. Cosmochim. Acta*, **31**, 407-430.

Tréhu, A. M. (1975) Depth versus age$^{1/2}$: a perspective on mid-ocean rises, *Earth Planet. Sci. Lett.*, **27**, 287-304.

Valentine, J. W., and Moores, E. (1972) Global tectonics and the fossil record, *J. Geol.*, **80**, 167-184.

Weertman, J. (1976) Milankovitch solar radiations and ice age sheet sizes, *Nature*, **261**, 17-20.

Some Perspectives of the Major Biogeochemical Cycles
Edited by Gene E. Likens
© 1981 SCOPE

CHAPTER 2

The Biogeochemical Nitrogen Cycle

T. ROSSWALL

SCOPE/UNEP International Nitrogen Unit,
Royal Swedish Academy of Sciences, Stockholm, Sweden

ABSTRACT

The biogeochemical nitrogen cycle is complex with nitrogen occurring in valence states from −3 to +5. Although abundant on earth, 96 per cent is found in the lithosphere and does not take part in the biogeochemical cycle. In terrestrial systems, only four per cent of the nitrogen occurs in biomass, while 96 per cent is found mainly in soil organic matter.

Nitrogen is an essential nutrient but nitrogen compounds are also potential toxicants. The hazard of unwanted side effects is increasing, i.e. through the increased use of nitrogen fertilizers. By the end of this century, man made additions of combined nitrogen to terrestrial ecosystems will be as large as the amount added through biological nitrogen fixation.

An understanding of the individual processes of the nitrogen biogeochemical cycle is needed before attempts can be made to quantify the nitrogen cycle of ecosystems, regions, or the earth. The inorganic nitrogen cycle, which was earlier considered relatively simple, is becoming more complex as our knowledge increases.

To establish a quantitative nitrogen budget for a system, we need extrapolation in time and space. Extrapolation in space is difficult due to the patchy distribution of anaerobic microsites. Extrapolation in time calls for a thorough understanding of the factors affecting process rates in order to quantify yearly rates based on a few measurements only.

The nitrogen cycle in an ecosystem is an important characteristic. The pattern of nitrogen cycling changes as an ecosystem develops to its climax state, and it is also affected by disturbances such as management practices.

A quantification and understanding of the nitrogen cycle in various types of typical ecosystem of the world will offer us the jigsaw pieces needed to construct an authoritative picture of the global nitrogen cycle, intrinsically linked to other major elements such as carbon, phosphorus and sulphur.

2.1 INTRODUCTION

Few elements are as complex and interesting as nitrogen. Firstly, this complexity is reflected in the highly intricate biogeochemical cycle, where nitrogen occurs in valence states from −3 to +5 and where many of the transformations are carried out by a few organisms only, at normal temperatures and pressures.

Table 2.1. Distribution of Nitrogen on Earth. All values in Tg (10^{12} g). Data from Sweeney et al. (1978)

	Amount of nitrogen (Tg)	Proportion (%)
Lithosphere	574 × 10^8	93.8
Atmosphere	38 × 10^8	6.2
Hydrosphere	0.23 × 10^8	0.04
Biosphere	0.009 × 10^8	0.001

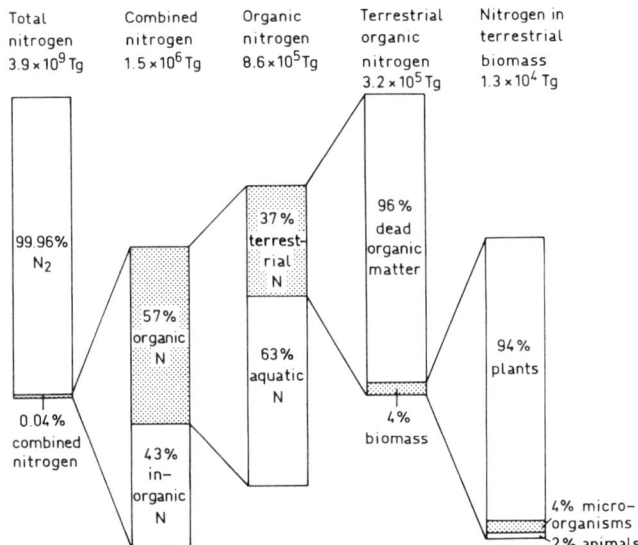

Figure 2.1 Distribution of nitrogen in the biosphere (Rosswall, 1979a)

Secondly, nitrogen is an element which is abundant on earth, but only a very small proportion of it enters into the biogeochemical nitrogen cycle at significant rates. Of the nitrogen found on earth, only 0.001 per cent occurs in the biosphere (Table 2.1). Most of the nitrogen is thus found in the lithosphere (especially in primary rocks of the mantle; see Stevenson, 1972). The atmosphere, where nitrogen occurs abundantly in its molecular form, still only contains six per cent of all nitrogen on earth. Of all nitrogen taking part in the biogeochemical nitrogen cycle only 0.04 per cent occurs in compounds potentially available to living organisms (Figure 2.1). In the terrestrial system only four per cent occurs in biomass, the remainder forming a large reservoir as dead organic matter (96 per cent). The unavailability of the 99.96 per cent occurring as nitrogen gas, combined with the major role played by nitrogen-containing substances in all forms of life, has caused nitrogen to be one of the key elements limiting the primary production on which man depends for his supply of food, fodder, fibre, and fuel.

Thirdly, the nitrogen cycle is easily manipulated by man, and it has been estimated that, by the end of this century, man-made additions of combined nitrogen will equal the amounts fixed annually through biological nitrogen fixation (Söderlund and Svensson, 1976). This increased addition from fertilizers, together with nitrogen oxides emitted into the atmosphere as a result of combustion, is undesirable since nitrogen compounds are a direct environmental hazard (Bolin and Arrhenius, 1977).

A SCOPE project on the global nitrogen cycle was started in 1974. It was terminated by the publication of a report quantifying the global cycle in detail not earlier attempted (Figure 2.2). The increased interest in the nitrogen cycle is reflected in the many global nitrogen budgets that have recently been published, e.g. Söderlund and Svensson (1976), Delwiche (1977), Hahn and Junge (1977), Sweeney et al. (1978), NAS (1978), and Bolin 1979. One of the main reasons behind this interest was the concern that nitrogen oxides may be acting as important catalytic agents regulating the thickness of the global stratospheric ozone shield. Our understanding of the nitrogen cycle is still, however, very rudimentary, and only by trying to understand the main factors regulating the many processes in the biogeochemical nitrogen cycle will it be possible to attain an improved quantitative picture of the global nitrogen cycle. An authoritative quantitative description is needed before we can quantify man's interventions in this cycle and their possible deleterious effects on the environment.

2.2 NITROGEN METABOLISM IN ECOLOGICAL TERMS

2.2.1 Nitrogen Assimilation

Plants and most microorganisms are dependent on ammonium or nitrate salts for growth, though they seem generally to prefer ammonium as a nitrogen source. Ammonium nitrogen can be metabolized by two different pathways—either through glutamate dehydrogenase (GDH) or through glutamine synthetase/glutamate synthase (GS/GOGAT) (Figure 2.3). Although GDH was previously considered the major enzyme involved in ammonium assimilation, it now seems clear that, especially at low ammonium concentrations, GS/GOGAT is the major enzyme system involved both in microorganisms and in plants (Miflin and Lea, 1977; Brown and Johnson, 1977); Lee and Stewart, 1978). This latter system has a much higher affinity for the substrate (K_m for GS: $1-2 \times 10^{-5}$ M; Lee and Stewart, 1978) than glutamate dehydrogenase ($K_m = 4 \times 10^{-3}$ M; Miflin and Lea, 1977). It should be noted that glutamine synthetase is ATP-dependent, whereas glutamate dehydrogenase is not. This system is thus probably the rate-limiting step, since ammonia uptake seems to be a passive process (Higinbotham, 1973).

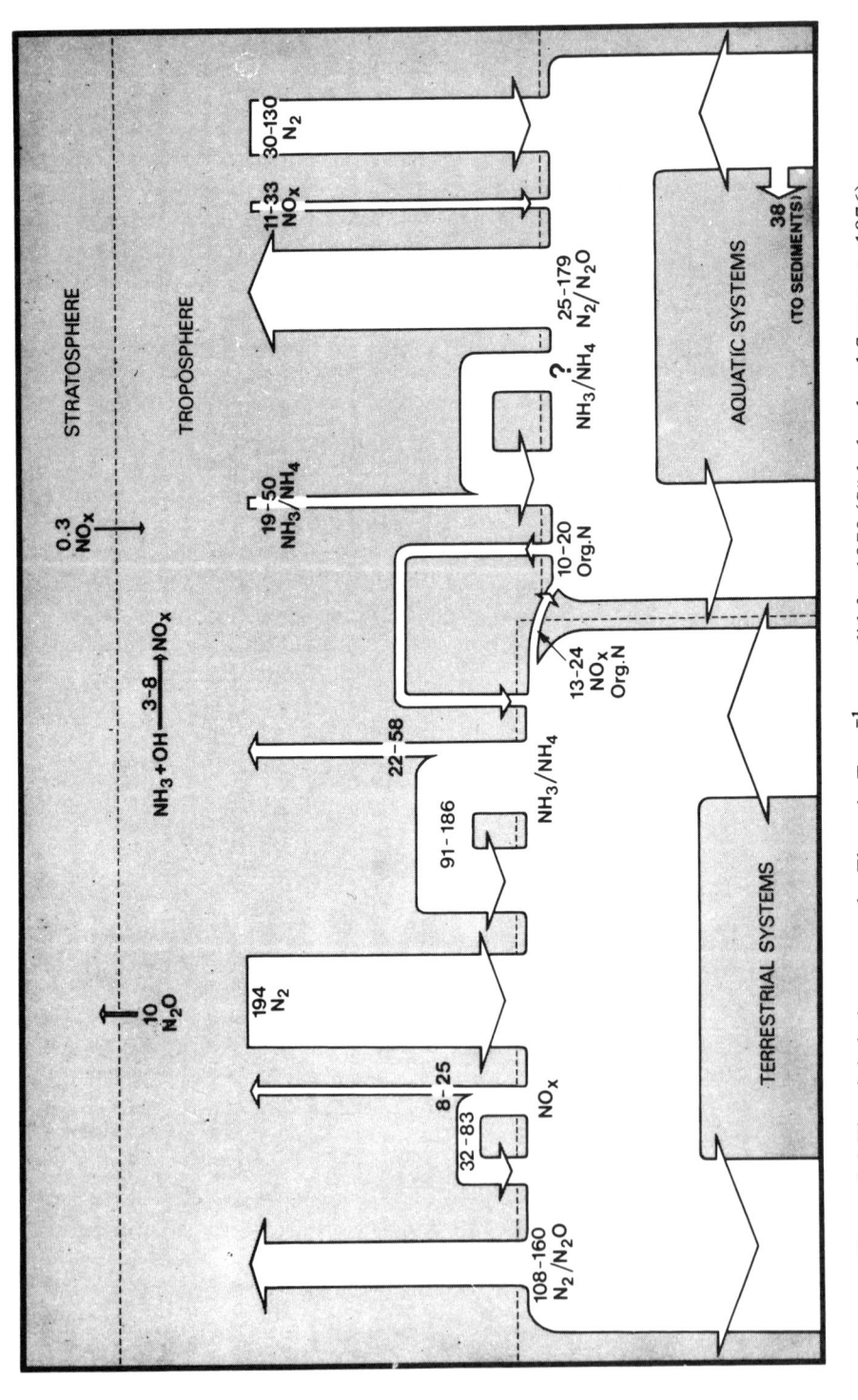

Figure 2.2 The global nitrogen cycle. Flows in Tg yr^{-1} are valid for 1970 (Söderlund and Svensson, 1976)

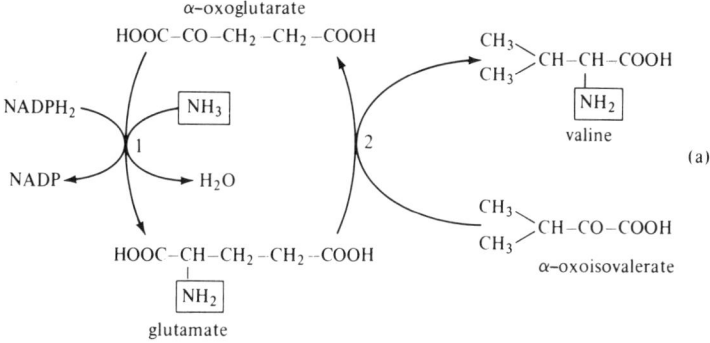

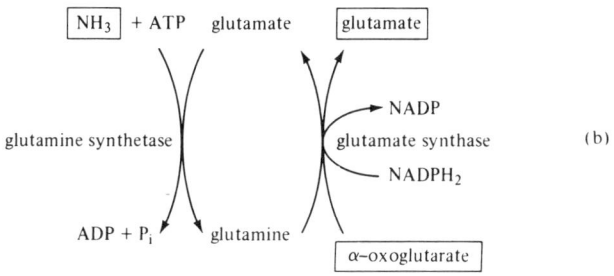

Figure 2.3 Assimilation of ammonia (a) by glutamate dehydrogenase (1) and subsequent transfer of the amino group by a transaminase (2); (b) by glutamine synthetase/glutamate synthase (GS/GOGAT. From Gottschalk (1979)

All plants, except certain bog species, are able to utilize nitrate as well as ammonium. If no other factors limit microbial growth, microorganisms will efficiently scavenge the surroundings for available ammonium nitrogen. Since ammonium oxidation through nitrification proceeds rapidly in most soils, nitrate is probably the most important nitrogen source for most plants. In contrast to ammonium uptake, plant root uptake of nitrate is dependent on a permease system (Lee and Stewart, 1978). The occurrence of nitrate reductase in the plant seems to give an indication of the rate of nitrate supply in the plant (Lee and Stewart, 1978). A comparison of net assimilation or net growth with nitrate reductase activity at different points in time could give an indication of the relative importance of ammonium and nitrate as the nitrogen source for plants at different times during the growing season.

Microorganisms generally prefer ammonium as a nitrogen source, and their ability to use nitrate is restricted. Of the 2500 genera of fungi described, only 20 have been reported to assimilate nitrate (Payne, 1973; Downey, 1978). The occurrence of nitrate assimilation in bacteria seems to be more common than in fungi, although it is in no way ubiquitous (Hall, 1978). Since the assimilatory nitrate reductase is repressed by ammonium (Gottschalk, 1979), the latter is the preferential nitrogen source for microorganisms.

2.2.2 Mineralization

Ammonium is liberated through mineralization, mainly by microorganisms, from organic compounds. Immobilization is the opposite process, whereby inorganic nitrogen is assimilated and built up into organic compounds. For soil microorganisms, the balance between mineralization and immobilization of ammonium nitrogen, i.e. positive or negative nitrogen net mineralization, is regulated primarily by the C/N ratio of the substrate (Parnas, 1975, 1976). Microbial biomass in soil has been estimated to contain four per cent nitrogen (Rosswall, 1976), and if the carbon content is 50 per cent, the microbial biomass has a C/N ratio of 12.5. When an organic substrate is broken down by microorganisms, the quality of the substrate will determine the relative proportion of carbon assimilated and respired. If the assimilatory efficiency is 40 per cent (Heal and MacLean, 1975), no net mineralization will occur if the C/N ratio is above 31. There are, however, very few data available on the assimilatory efficiency of microorganisms growing in soil, and this efficiency is very critical in determining the C/N ratio below which net mineralization occurs (Figure 2.4). However, one cannot judge the mineralization capacities of soils from their C/N ratios alone. In systems which accumulated organic matter, the C/N ratio is generally about 30 (Parsons and Tinsley, 1975). In tundra peat it has been observed to be as high as 48 in the top 10 cm (Rosswall *et al.*, 1975), where most of the net mineralization occurs. Scots pine needle litter from a coniferous forest on glacial sediment in central Sweden originally had a C/N ratio as high as 120, whereas at the time when net mineralization started it was 80 (Staaf and Berg, 1977). If microbial biomass contains four per cent N, carbon/nitrogen ratios of 48 and 120 would require assimilatory efficiencies as low as 23 and 10 per cent respectively for net mineralization to occur (Figure 2.4).

One of the main difficulties in using such a crude concept as the C/N ratio of the substrate for determining whether mineralization or immobilization occurs relates to the forms of nitrogen in organic matter. Part of the nitrogen, especially that bound in proteins, is often easily mineralized, but another part, bound to the lignin fraction, is very resistant to mineralization. Such differences must be kept in mind in any attempt to describe net nitrogen mineralization.

Soil animals may play an important role in regulating nitrogen mineralization in at least two ways. The effect of faunal grazing on microorganisms may increase mineralization rates (Rosswall *et al.*, 1977; Coleman *et al.*, 1977). It also seems that soil invertebrates may play an important role through their excretion of significant amounts of simple nitrogenous substances, such as uric acids, urea, and also ammonia (Table 2.2). They may thus play a more important role in nitrogen mineralization than hitherto assumed.

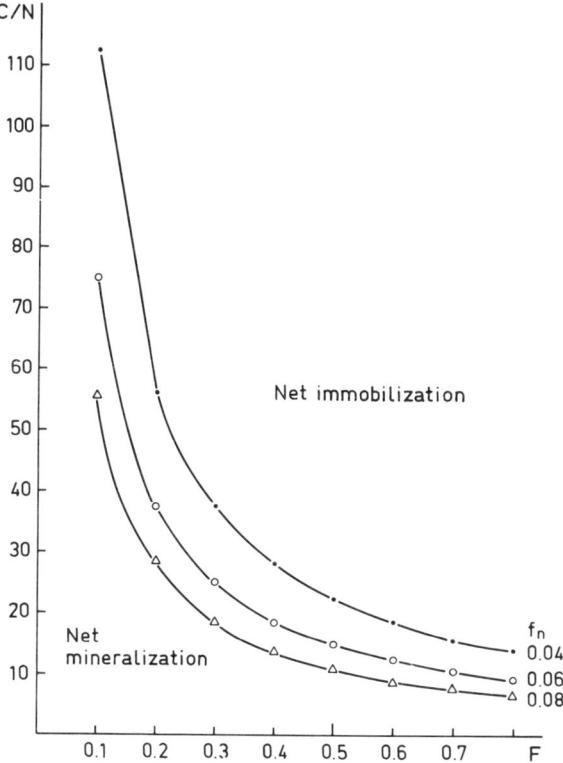

Figure 2.4 The dependence of net mineralization versus immobilization on the carbon/nitrogen ratio of the substrate, the assimilatory efficiency of the decomposers (F), and the nitrogen content of microbial biomass (f_n)

2.2.3 Nitrification-Denitrification

Nitrogen also plays an important role in generating energy in certain groups of microorganisms. Autotrophic nitrifying bacteria obtain energy from the oxidation of ammonium or nitrite, while denitrifying bacteria use nitrate, nitrite, or nitrous oxide as terminal electron acceptors during the oxidation of organic substrates. Nitrate can also be used in a fermentative process by *Clostridium perfringens* (Hasan and Hall, 1975), although since the process uses an inorganic electron acceptor, it is by definition a respiratory process. The amounts of energy obtained by these processes are listed in Table 2.3.

There seem to be very close couplings between nitrification and denitrification (Figure 2.5). It has been postulated that the nitroxyl radical HNO is a possible intermediate in the oxidation of hydroxylamine to nitrite by *Nitrosomonas* (Lees,

Table 2.2 Examples of Nitrogenous Substances Excreted by Soil Animals. The Figures are Percentages of Total Nitrogen Excretion (Persson, 1980)

	NH_3	Urea	Allantoin	Uric acid	Amino-N	Other
PROTOZOA						
Ciliates	++				+	
NEMATODES						
Ditylenchus	39				28	
EARTHWORMS						
Lumbricus terrestris (well fed)	56	42		1.5		
L. terrestris (starving)	19	81				
WOODLICE						
Oniscus asellus	47			5	6	38
INSECTS						
Forficula (earwigs)				100		
Heteroptera			60–100	0–40		
Lucilia-larvae (blowfly)	93		2	5		
Bibio marci (terrestrial diptera larvae)			55	25		
SNAILS						
Helix pomatia				43		Xanthin 38 Guanin 18

1954). There is still no direct evidence for this, however (Nicholas, 1978). The presence of the nitroxyl radical would also explain the possibilities for N_2O production during nitrification, since, especially under anaerobic conditions, there would be a spontaneous (chemical) decomposition of [HNO] to N_2O (Anderson, 1964).

Nitroxyl also has been suggested by Kluyver and Verhoeven (1954) to be an intermediate in nitrite reduction by denitrifying bacteria, and since a dimerization step is entailed, this radical would seem to be a probable intermediate (Delwiche and Bryan, 1976).

Nitrosomonas also appears to reduce nitrite under anaerobic conditions (Ritchie and Nicholas, 1972). From an ecological viewpoint, there may be several advantages (Rosswall, 1980):

(i) nitrite may be used as a terminal electron acceptor during brief periods of anaerobiosis;

(ii) nitrite reduction may offer a detoxifying mechanism if nitrite concentrations become too high; and

(iii) under certain circumstances nitrite reduction may minimize possible competition from *Nitrobacter* which is dependent on nitrite for growth.

Table 2.3 Free Energy Changes (kJ mol^{-1}) in Inorganic Nitrogen Metabolism Reactions (from Zumft and Cárdenas, 1979; Delwiche, 1977)

	$\Delta G'_0$ (kJ/mol)
Nitrate respiration *Escherichia coli* $NO_3^- + H_2 \rightarrow NO_2^- + H_2O$	-161
Denitrification *Pseudomonas aeruginosa* $2 NO_3^- + 2 H^+ + 5 H_2 \rightarrow N_2(g) + 6 H_2O$	-1121
Other possible reactions	
$N_2O(g) + H_2 \rightarrow N_2(g) + H_2O$	-340
$NO_2^- + \frac{1}{2} H_2 + H^+ \rightarrow NO(g) + H_2O$	-76
$2 NO(g) + H_2 \rightarrow N_2O(g) + H_2O$	-306
$2 NO_2^- + 2H^+ + 2 H_2 \rightarrow N_2O(g) + 3 H_2O$	-459
Nitrate reduction $NO_3^- + 2 H^+ + H_2O \rightarrow NH_4^+ + 2 O_2$	$+348$
Nitrate fermentation *Clostridium perfringens* $NO_3^- + 2 H^+ + 4 H_2 \rightarrow NH_4^+ + 3 H_2O$	-591
Nitrification *Nitrosomonas* $NH_4^+ + \frac{1}{2} O_2 \rightarrow NH_2OH + H^+$	$+15$
$NH_2OH + O_2 \rightarrow NO_2^- + H_2O + H^+$	-289
Nitrobacter $NO_2^- + \frac{1}{2} O_2 \rightarrow NO_3^-$	-77

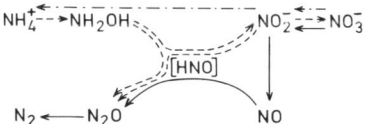

Figure 2.5 The inorganic nitrogen cycle in soil

Certain heterotrophic microorganisms may also oxidize ammonium to nitrate (Focht and Verstraete, 1977). Heterotrophic nitrification may be important in view of the possible ecological significance of some of the products identified. Hydroxamic acids may be important growth factors, possibly linked to iron uptake, and several of the identified metabolic products from heterotrophic nitrification are biocidal (Focht and Verstraete, 1977).

Another interesting feature in the metabolism of inorganic nitrogen relates to the inhibitory effect of acetylene on several of the enzyme systems involved, viz.

nitrogenase (Dilworth, 1966), ammonium oxidase (Suzuki, 1978; Hynes and Knowles, 1978), and nitrous oxide reductase (Balderston *et al.*, 1976; Yoshinari and Knowles, 1976). The effect on nitrification may complicate the use of the acetylene inhibition technique for determining denitrification in systems that are dependent on nitrate production by nitrifying bacteria (Hynes and Knowles, 1978).

2.2.4 Nitrogen Fixation

Interest in nitrogen fixation has increased in recent years for several reasons. From a practical viewpoint, it was largely triggered by the tripling in the price of nitrogen fertilizers after 1974, and by the difficulties—especially for farmers in the less developed countries—of supplying nitrogen to the soil in sufficient amounts to increase crop production. Much attention was accordingly given to optimizing the use of biological nitrogen fixation.

The use of *Rhizobium* for inoculating legume crops constitutes the most promising approach and has been widely used for many years. Significant research projects are being carried out in a large number of countries to optimize biological nitrogen-fixation using *Rhizobium*. In tropical ecosystems the contribution of indigenous legumes is often more important that that of cultivated species, such as soy bean, peanut, cowpea and chick-pea (A. Foury, 1950; cited by Franco, 1978). Specialized regional microbiological resource centres (MIRCENs) devoted to the use of biological nitrogen fixation have been set up in Kenya and Brazil under the sponsorship of UNEP and Unesco (Rosswall, 1979b). Important research efforts also are being made by the international agricultural research institutes, e.g. International Institute for Tropical Agriculture (IITA) in Nigeria and the International Crop Research Institute for the Semi-arid Tropics (ICRISAT) in India. Strain selection from native legumes will be important, especially the isolation of *Rhizobium* strains with an efficient hydrogenase, since strains which can recycle the hydrogen formed during nitrogen fixation have been shown to be more efficient in fixing nitrogen (Albrecht *et al.*, 1979).

Attention has also been focused recently on the possibilities of associative symbioses between microorganisms and plant roots, the most important of which seems to be the association of *Azospirillum* with sorghum, maize (von Bülow and Döbereiner, 1975), and sugar cane (Ruschel *et al.*, 1978). The association of *Azotobacter paspali* and *Paspalum notatum* is another such association that has received increased attention (Döbereiner, 1970, 1977). The possibility that *Azospirillum* may denitrify and thus fix nitrogen under anaerobic conditions, with nitrate as a terminal electron acceptor, is another interesting recent finding (Neyra and Van Berkum, 1977), which is a futher important link in the coupling of inorganic nitrogen metabolism in bacteria, as discussed above.

The aquatic water fern *Azolla*, together with the blue-green alga *Anabaena azollae*, provides another example of a nitrogen-fixing association that can fix substantial amounts of nitrogen and be of agronomic significance (Moore, 1969). It is especially important in rice fields (Talley et al., 1977). The recent isolation of the cyanobacterium responsible for the nitrogen fixation (Newton and Herman, 1979) will be important in further attempts to investigate—and ultimately select—strains which are efficient in the symbiosis. An *Azolla* species which excretes ammonium during the entire growing season seems to be especially promising for agronomic use (R. C. Vallentine, personal communication).

Non-legume symbiosis with actinomycetes of the genus *Frankia* occurs in several temperate species, such as *Alnus glutinosa, Hippophaë rhamnoides* and *Myrica gale* (see Bond, 1967, for a review). *Alnus* species, for example, may become important in short-term, rotation forestry for energy production (Zavitkovski et al., 1979). *Myrica gale* is a common bog species in Europe and N. America and could be important in the nitrogen economy of such bog ecosystems (Sprent et al., 1978). *Myrica* can perhaps be used in the early stages of forest plantation after drainage. Recent reports of the successful isolation of *Frankia* from *Alnus* nodules and its growth in pure culture in laboratory media (Baker and Torrey, 1979) constitute an important breakthrough in the attempts to understand this symbiosis.

Other nitrogen-fixing associations include leaf nodules of the angiosperm *Gunnera*, cycad-bluegreen algae root-nodule symbiosis, liverworts and lichens (see Stewart, 1977, for review). New nitrogen-fixing species are, however, still being discovered, such as the *Rhizobium* symbiosis with the legume *Aeschynomena indica*, forming large numbers of stem nodules fixing nitrogen (Yatazawa and Yoshida, 1979), and the herb *Datisca cannabina*, which has been shown to form root nodules of the *Alnus* type (Chaudhary, 1979). It is likely that a large number of nitrogen-fixing species is as yet unidentified, some of which may prove useful in agriculture and forestry. For tropical ecosystems, it seems especially important to consider the use of legume trees in crop-rotation practices and multiple cropping systems, and their importance in savanna ecosystems has been stressed (Rosswall and Vitousek, in press).

Interest has also been shown in the interaction between nitrogen-fixing bacteria and animals, one example being nitrogen fixation in the gut of termites (Breznak et al., 1973). The bacterium responsible, *Citrobacter freundii*, has been isolated from termites, and it may be important in supplying combined nitrogen to the animals, which feed mainly on carbohydrate-rich material with a low nitrogen content (French et al., 1976).

In the more distant future, the possibilities of introducing the genes (*nif* genes) for nitrogen fixation into new hosts that can be used for practical purposes is a fascinating prospect (see, for example, Postgate, 1977; Klingmüller, 1979), even if there are still very large difficulties.

2.3 A MICRONICHE CONCEPT OF NITROGEN CYCLING

Most biological nitrogen transformations are brought about by microorganisms, these processes being regulated both qualitatively and quantitatively by a large

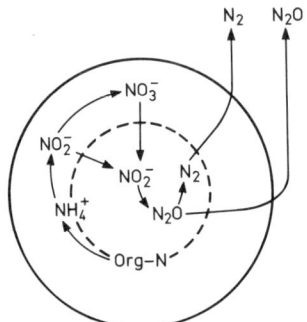

Figure 2.6 A schematic soil crumb with an aerobic outer sector in which nitrification can take place and an inner anaerobic centre from which nitrogen may be lost through denitrification (Knowles, 1978)

number of factors. The qualitative aspects are primarily regulated by oxygen and energy sources, while the quantitative aspects are regulated mainly by temperature and moisture. For example, soil water has a dual function. Moisture is a rate-limiting factor for all biological processes. At moisture contents below 20 per cent, activity is generally low, while at higher moisture contents there is a rapid increase in activity; the process often reaches a maximal rate at about 80 per cent, after which it starts to decline (e.g. Bunnell and Tait, 1974). Soil moisture regulates process rates and is thus an important quantitative factor. However, the often observed decreased rate at very high moisture contents reflects a qualitative change rather than a direct quantitative change. At increased water contents, oxygen diffusion becomes limiting and the microbial metabolism switches from aerobic to anaerobic. Water then acts as a regulator of the quality and only secondarily of the quantity, since anaerobic processes are generally slower than aerobic.

Water has a profound effect in the soil and makes generalizations of process rates (expressed as grams of nitrogen transformed per square metre) very difficult. As an example, let us consider the inorganic nitrogen metabolism of microorganisms. A schematic representation of a soil crumb is shown in Figure 2.6, in which organic nitrogen is mineralized most rapidly in the aerobic zone; after nitrification, nitrate can diffuse into the anaerobic centre, where it can be denitrified, and gaseous nitrogen products are lost to the system. It has been estimated that most soil crumbs with a diameter of more than 3 mm have anaerobic centres (Currie, 1961; Greenwood, 1969). It is also possible that *Nitrosomonas* bacteria can briefly switch to anaerobic metabolism, using nitrite as an electron acceptor, as previously discussed. Nitrogen-fixing aerobic bacteria may be able to use nitrate as a terminal electron acceptor and continue to fix nitrogen under anaerobic conditions, since it has been

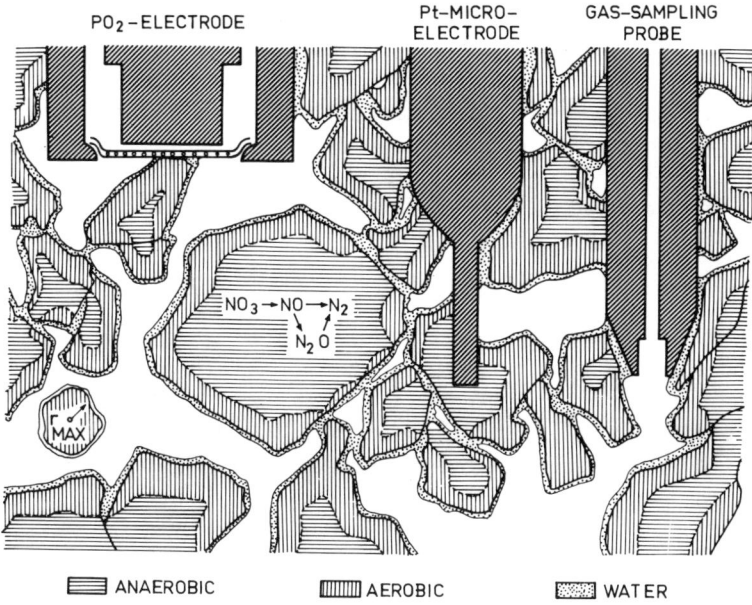

Figure 2.7 Schematic drawing showing the various sensors used to assess the extent of anaerobiosis in soil in relation to soil structural elements (Flühler et al., 1976)

shown that *Rhizobium* (Zablotowicz et al., 1978) and *Azospirillum* (Eskew et al., 1977; Neyra and Van Berkum, 1977) can denitrify. It is possible that the nitrous oxide production observed from the nitrogen-fixing *Klebsiella pneumoniae* (Yoshida and Alexander, 1970) will be shown to be an end-product of denitrification (Rosswall, 1978), since it does not seem as if N_2O can be produced during assimilatory nitrite reduction (Losada, 1975/76).

The occurrence of anaerobic microsites is probably also the reason for the observed denitrification rates in pelagic invertebrate fecal pellets (Söderlund and Svensson, 1976; Wilson, 1978).

The difficulties of determining the aerobic versus anaerobic activities in soils are linked to the difficulty of determining the extent of anaerobiosis. A schematic representation of a soil system with the various sensors which can be used for determining the aeration status of a soil, viz., an oxygen electrode, a redox electrode and a gas sampling probe, is shown in Figure 2.7. Pt-microelectrodes with a diameter of less than one micron have been developed for use in live tissues (Lübbers, 1969), but a very large number of determinations must be made to determine the extent of anaerobiosis if such microsites are few (Flüher et al., 1976).

The difficulties of determining the different processes in the biogeochemical nitrogen cycle even on a square-metre basis are obvious. The difficulty of extrapolation, however, not only relates to space but also to time. Single or even repeated measurements of, nitrogen fixation, for example, are very difficult to extrapolate to a yearly basis (Burris, in press). Soil bacteria show very rapid fluctuations in numbers just over a few days, especially after rainfall (Clarholm and Rosswall, 1980), a fact which provides another indication of the difficulty of extrapolating monthly rate determinations to yearly rates. Frequent measurements must be made at the time of rainfall or irrigation as well as during high and low temperature regimes before realistic yearly rates can be calculated.

2.4 ECOSYSTEM NITROGEN CYCLES

The pattern of nitrogen cycling, together with that of the cycling of other nutrients, is an important factor in describing the functioning of ecosystems. Odum (1969) postulated how nutrient cycling relates to the development of ecosystems. Detailed studies have been made on nitrogen cycling in different ecosystems, but these have generally been made assuming steady-state conditions. Only recently has attention been given to non-steady-state conditions such as comparisons of different successional stages, and impact of disturbances on ecosystem nitrogen cycles.

The concept of an increasing retention of available nitrogen during the successional development of a terrestrial ecosystem has been interpreted in a number of ways. Haines (1977) suggested the following possible hypotheses:

(i) roots in succeeding stages become more efficient in the uptake of dissolved and exchangeable nutrients;
(ii) increased soil organic matter in succeeding stages retains an increasing proportion of dissolved and exchangeable nutrients;
(iii) a smaller proportion of the total nutrient pool of succeeding stages is lost from the system; and
(iv) in succeeding stages, a smaller proportion of the nutrient pool of the system is in flux at any given time.

It seems as if nitrate uptake decreases while ammonia uptake increases with succession (Haines, 1977). The possibility of an allelopathic effect of root exudates of climax species has been advanced as a possible explanation for the observed decrease in nitrate uptake (Rice and Pancholy, 1972, 1973), but there is only circumstantial evidence that this is the case. Only by understanding what factors affect the various processes in the nitrogen cycle will it be possible to evaluate the merits of such an assumption.

The fact that nitrate production can be accelerated in ecosystems subject to destructive disturbance was noted early on by Hesselman (1917), and a number of studies on the magnitude of nitrogen losses following disturbance have been made; one of the major studies being carried out is that in the Hubbard Brook Experimental Forest (Likens *et al.*, 1977; Bormann and Likens, 1979). The leaching of nitrogen from ecosystems following disturbance was recently investigated in a number of forest ecosystems in the USA (Vitousek *et al.*, 1979). It has been suggested that the concepts of resistance and resilience could be used in describing the

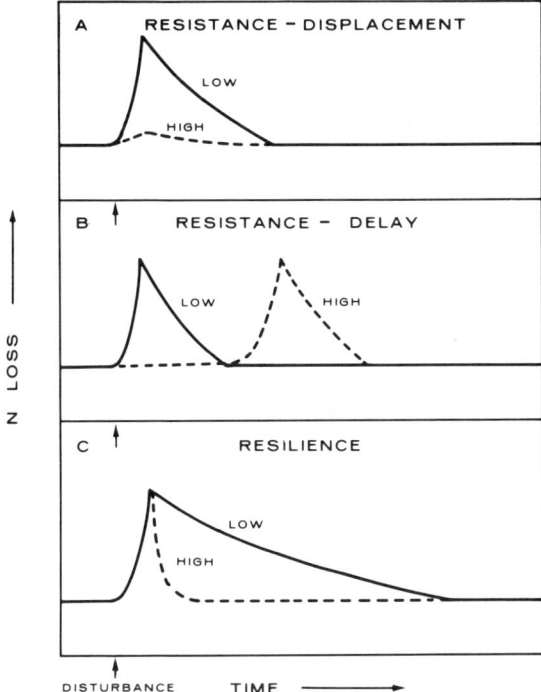

Figure 2.8 Relations between resistance and resilience of ecosystems to disturbance and subsequent nitrogen losses (Vitousek, 1980)

effects on nitrogen leaching from ecosystems after disturbance. Resistance is the ability of an ecosystem to withstand perturbations, while resilience concerns the speed with which the system returns to its original condition. With regard to the resistance of ecosystems to nitrogen losses by leaching, Vitousek (1980) suggested that these are of two types, one related to the total amount of nitrogen leached (displacement) and the other the resistance in time (delay). Using these terms, Vitousek proposed the following factors to be related to the mentioned group of responses (Figure 2.8):

(i) resistance to displacement, which is related to rates of nitrogen cycling before perturbation;
(ii) delay, which is caused by nitrogen immobilization; and
(iii) resilience, which is related to the rate of re-establishment of plant nitrogen uptake.

Most attention has been given the quantification of inputs and outputs of nitrogen in studies of the biogeochemical nitrogen cycle. It should be noted,

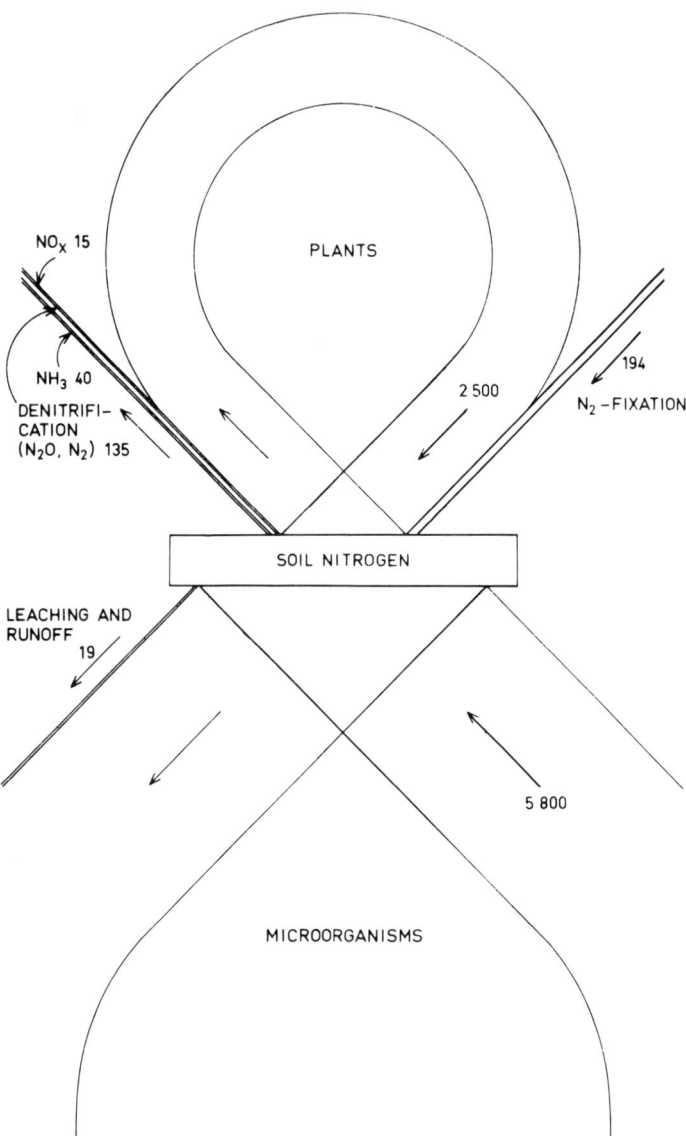

Figure 2.9 Annual transfers of nitrogen on a global basis between primary producers and microorganisms as compared to flows to and from the terrestrial system. All flows in Tg yr^{-1} and storages in Tg (Rosswall, 1976)

however, that the inputs and outputs of the nitrogen circulating in the vegetation-microorganism-soil system annually (Figure 2.9). To understand what factors regulate transfers of nitrogen to and from ecosystems, further detailed knowledge of the processes involved in nitrogen transformations within the system is essential.

Data on ecosystem nitrogen budgets and an evaluation of the functional characteristics of the cycles are needed as important jigsaw pieces in the efforts to produce an authoritative and quantitative description of the global nitrogen cycle. Such an understanding is vital if we are to evaluate man's impact on the nitrogen cycle and the consequences that any disturbance might have.

2.5 THE GLOBAL NITROGEN CYCLE

Some recent global nitrogen budgets are summarized in Table 2.4 and it is evident that large uncertainties exist (see also Chapter 6). For example, to what extent does nitrification contribute to the global production of nitrous oxides, and do the oceans and terrestrial systems act both as a sink and source for nitrous oxide? We are slowly developing an ecological theory regarding nitrogen cycling in ecosystems and, based on an understanding of the individual processes, efforts are being made to generalize the observed strategies that ecosystems have developed. Knowledge of the theory of microbiological nitrogen transformations and ecosystem theory will be indispensable when attempts are made to forecast the effect of management practices on the biogeochemical nitrogen cycle.

The scale on which qualitative data on nitrogen cycles is needed depends on the question under consideration. An interpretation of nitrogen retention must relate to the competition between microorganisms and plant roots to utilize ammonium-N or nitrate-N. A microniche concept is then needed. Developing an optimal agroecosystem management practice with regard to crop uptake of added nitrogen fertilizers calls for an understanding of the biogeochemical nitrogen cycle at the ecosystem level. An evaluation of the possible effect of nitrogen oxides on the stratospheric ozone layer requires a quantification of the global nitrogen budget.

Only by combining basic laboratory studies with observations of nitrogen cycling dynamics in representative ecosystems—for example, as parts of the MAB programme—and by trying to integrate these data into a global quantitative model, will it be possible to meet the challenge for supplying the data badly needed for a cost-benefit analysis of fertilizer use and environmental risk assessments related to the increasing occurrence of nitrogen compounds as pollutants.

2.6 INTERACTIONS BETWEEN NITROGEN AND OTHER BIOGEOCHEMICAL CYCLES

The interactions between biogeochemical cycles are covered by other papers in this volume. It should be stressed that the biogeochemical nitrogen cycle should not be viewed in isolation, as there are close couplings with cycles of, for example, water, carbon, phosphorus, and sulphur.

As mentioned above, the water cycle is instrumental in determining the fate of

Table 2.4 Estimates of Global Nitrogen Transfers

	Eriksson (1959)	Robinson and Robbins (1970)	Delwiche (1970)	Burns and Hardy (1975)	Söderlund and Svensson (1976)	McElroy et al. (1976)	CAST (1976)
Biological fixation: land	104	118	44	139	139	170	149
ocean	n.d.	12	10	36	30–130	10	1
Atmospheric fixation (lightning)	n.d.	n.d.	8	10	?	10	10
Industrial fixation	n.d.	20	30	30	36[1]	40	57[2]
Combustion	15	19	n.d.	20	19[1]	40	20
Fires	n.d.	n.d.	n.d.	n.d.	n.d.	n.d.	n.d.
Biogenic NO_x production	n.d.	234	n.d.	n.d.	21–89	n.d.	n.d.
Denitrification: land	65	n.d.	43	140	107–161	243	70–100
oceans	87	n.d.	40	70	25–179	106	70–100
Ammonia volatilization	99	957	n.d.	165	113–244	150	n.d.
Dry deposition NH_3/NH_4^+	n.d.	175	⎫	n.d.	72–151	⎫	⎫
Wet	99	796	⎬ n.d.	140	38–85	⎬ 220	⎬ n.d.
Dry deposition NO_x	n.d.	22	⎪	n.d.	25–70	⎪	⎪
Wet	48	83	⎭	60	18–46	⎭	⎭
Dry deposition Org. –N	n.d.	n.d.		n.d.	?		
Wet	36	n.d.		n.d.	10–100		
River runoff NH_4^+	—	⎫	⎫	⎫	<1	⎫	⎫
NO_3^-	21	⎬ 13	⎬ 30	⎬ 15	5–11	⎬ 20	⎬ n.d.
Org. –N	—	⎭	⎭	⎭	8–13	⎭	⎭
N_2O sink: land	n.d.	353	n.d.	n.d.	n.d.	0	n.d.
oceans	n.d.					50	

Table 2.4 (Continued)

	Delwiche and Likens (1977) and Delwiche (1977)	Liu et al. (1977)	Hahn and Junge (1975)	Sweeney et al. (1978)	NAS (1978)	Bolin (1979)
Biological fixation: land	99	200	180	100	139	140
ocean	30	40	85	15–90	100	20–120
Atmospheric fixation (lightning)	7	10	n.d.	0.5–3	30	?
Industrial fixation	40	40	40	35	70[3]	40
Combustion	18	20	n.d.	15	21	20
Fires	50	n.d.	n.d.	n.d.	10–200	n.d.
Biogenic NO_x production	—	n.d.	?	n.d.	22–66	20–90
Denitrification: land	120	140	150	90	197–390	63–245
oceans	40	130	165	50–125	0–120	35–330
Ammonia volatilization	75	190	170	15[5]	36–90	110–250
Dry deposition NH_3/NH_4^+ Wet	79	} 220[4]	} 200	} n.d.	} 83–242	} 110–240
Dry deposition NO_x Wet	34					40–110
Dry deposition Org.–N Wet	n.d.					n.d.
River runoff NH_4^+	} 35	} 30	} 40	} 30	} 18	} 25–35
NO_3^-						
Org.–N						
N_2O sink: land	n.d.	n.d.	n.d.	n.d.	n.d.	n.d.
oceans						

n.d. = no data given; [1] data for 1970; [2] data for 1974; [3] data for 1976; [4] precipitation only; [5] net transfer from land to oceans.

nitrogen in an ecosystem. Water regulates the rates of the processes involved in nitrogen transformations, it acts as a diffusion barrier for oxygen creating anaerobic conditions and it transports nitrogen to and from the ecosystem.

The bulk of soil nitrogen occurs in organic matter and mineralization of organic carbon leads to a release of ammonia nitrogen, although the relative proportions of carbon and nitrogen play an important regulatory role in net nitrogen mineralization as discussed above. When nitrogen fertilizers are added to a crop, photosynthesis and carbon immobilization ususally increase, but fertilization could also lead to increased soil organic matter decomposition with subsequent release of carbon dioxide.

Specific examples of the effect of phosphorus on nitrogen transformations are many. Nitrogen fixation and nitrification seem to be particularly sensitive to the availability of soil phosphorus. In P deficient soils, nitrogen fixation by legumes is increased considerably if the legume has an efficient VA-mycorrhizal symbiont, through which P uptake can improve. (e.g. Crush, 1974; Daft and El-Giahmi, 1974; Sanni, 1976). In addition to the discussion above on the regulation of nitrification rates, Pancholy suggested that phosphorus deficiency was the main reason for restricted nitrification rates observed in phosphorus-poor savanna soils.

Both nitrogen and sulphur compounds can acidify the environment and such changes in pH will affect nitrogen cycling processes (Helyar, 1976). It is important to quantify the relative contributions of nitrogen and sulphur compounds to the possible acidification of soils and the fate of the acidifying compounds as they relate to the acid-base status. Nitrogen fixation and nitrification are particularly sensitive to reduced pH in soils.

2.7 REFERENCES

Albrecht, S. L., Maier, R. J., Hanus, F. J., Russell, S. A., Emerich, D. W. and Evans, H. J. (1979) Hydrogenase in *Rhizobium japonicum* increases nitrogen fixation by nodulated soybeans, *Science,* **203**, 1255-1257.

Anderson, J. H. (1964) The metabolism of hydroxylamine to nitrite by *Nitrosomonas, Biochem. J.,* **91**, 8-17.

Baker, D. and Torrey, J. G. (1979) The isolation and cultivation of actinomycetous root nodule endophytes, in Gordon, J. C., Wheeler, C. T. and Perry, D. A. (1979) *Symbiotic Nitrogen Fixation in the Management of Temperate Forests,* Corvallis, Oregon: Forest Research Laboratory, Oregon State University, 38-56.

Balderston, W. L., Sherr, B. and Payne, W. J. (1976) Blockage by acetylene of nitrous oxide reduction in *Pseudomonas perfectomarinus, Appl. Environ. Microbiol.,* **31**, 504-508.

Berg, B. (1978) *Decomposition of needle litter in a 120-year-old Scots pine (Pinus silvestris) stand at Ivantjärnsheden,* Swedish Coniferous Forest Project Technical Report 80.

Bolin, B. and Arrhenius, E. (eds) (1977) Nitrogen—An essential life factor and a growing environmental hazard, Report from Nobel symposium No. 38, *Ambio* **6**, 96-105.

Bolin, B. (1979) On the role of the atmosphere in biogeochemical cycles, *Quart. J. R. Met. Soc.,* **105**, 25-42.

Bond, G. (1967) Fixation of nitrogen by higher plants other than legumes, *Ann. Rev. Plant Physiol.,* **18**, 107–126.

Bormann, F. H. and Likens, G. E. (1979) *Pattern and Process in a Forested Ecosystem,* New York, Heidelberg, Berlin, Springer-Verlag.

Breznak, J. A., Brill, W. J., Mertins, J. W. and Coppel, H. C. (1973) Nitrogen fixation in termites, *Nature,* **244**, 577–580.

Brown, C. M. and Johnson, B. (1977) Inorganic nitrogen assimilation in aquatic microorganisms, *Adv. Aquatic Microbiol.* **1**, 49–114.

Bunnell, F. L. and Tait, D. E. N. (1974) Mathematical simulation models of decomposition processes, in Holding, A. J., Heal, O. W., MacLean, S. F. and Flanagan, P. W. (eds) *Soil Organisms and Decomposition in Tundra,* Stockholm, Tundra Biome Steering Committee, 207–225.

Burns, R. C. and Hardy, R. W. F. (1975) *Nitrogen Fixation in Bacteria and Higher Plants,* Berlin, Heidelberg, New York, Springer-Verlag.

Burris, R. H. (in press) The global nitrogen budget—science or seance, V, in Newton, W. E. and Orme-Johnson, W. (eds) *Nitrogen Fixation,* Baltimore, University Park Press, Volume 1, 7–16.

von Bülow, J. F. W. and Döbereiner, J. (1975) Potential for nitrogen fixation in maize genotypes in Brazil, *Proc. Nat. Acad. Sci. USA,* **72**, 2389–2393.

CAST (1976) *Effect of Increased Nitrogen Fixation on Stratospheric Ozone,* Council for Agricultural Science and Technology Report No. 53, Ames. Iowa, Iowa State University.

Chaudhary, A. H. (1979) Nitrogen-fixing root nodules in *Datisca cannabina* L., *Plant Soil,* **51**, 163–165.

Clarholm, M. and Rosswall, T. (1980) Biomass and turnover of bacteria in a forest soil and a peat, *Soil Biol. Biochem.,* **12** (in press).

Coleman, D. C., Cole, C. V., Anderson, R. V., Blaha, M., Campion, M. K., Clarholm, M., Elliott, E. T., Hunt, H. W., Shaefer, B. and Sinclair, J. (1977) An analysis of rhizosphere-saprophage interactions in terrestrial ecosystems, in Lohm, U. and Persson, T. (eds) Soil organisms as components of ecosystems, *Ecol. Bull. (Stockholm),* **25**, 299–309.

Crush, J. R. (1974) Plant growth responses to vesicular-arbuscular mycorrhiza, VII, Growth and nodulation of some herbage legumes, *New Phytol.,* **73**, 743–749.

Currie, J. A. (1961) Gaseous diffusion in the aeration of aggregated soils, *Soil Sci.,* **92**, 40–45.

Daft, M. J. and El-Giahmi, A. A. (1974) Effect on endogone mycorrhiza on plant growth, VII, Influence of infection on the growth and nodulation in French Bean (*Phaseolus vulgaris*), *New Phytol.,* **73**, 1139–1147.

Delwiche, C. C. (1970) The nitrogen cycle, *Scientific American,* **223(3)**, 137–146.

Delwiche, C. C. (1977) Energy relations in the global nitrogen cycle, *Ambio,* **6**, 106–111.

Delwiche, C. C. and Bryan, B. A. (1976) Denitrification, *Ann. Rev. Microbiol.,* **30**, 241–262.

Delwiche, C. C. and Likens, G. E. (1977) Biological response to fossil fuel combustion products, in Stumm, W. (ed.) *Global Chemical Cycles and Their Alterations by Man,* Berlin, Dahlem Konferenzen, 73–88.

Dilworth, M. J. (1966) Acetylene reduction by nitrogen-fixing preparations from *Clostridium pasteurianum, Biochim. Biophys. Acta,* **127**, 285–294.

Döbereiner, J. (1970) Further research on *Azotobacter paspali* and its variety specific occurrence in the rhizosphere of *Paspalum notatum* Flugge, *Zbl. Bakt. Parasitenk. II,* **124**, 224–230.

Döbereiner, J. (1977) Biological nitrogen fixation in tropical grasses—possibilities for partial replacement of mineral N fertilizers, *Ambio,* **6**, 174–177.

Downey, R. J. (1978) Control of fungal nitrate reduction, in *Microbiology-1978*, Washington, D.C., American Society for Microbiology, 320-323.

Eriksson, E. (1959) Atmospheric chemistry, *Sv. Kem. Tidskr.*, **71**, 15-32 (in Swedish).

Eskew, D. L., Focht, D. D. and Ting, I. P. (1977) Nitrogen fixation, denitrification, and pleomorphic growth in a highly pigmented *Spirillum lipoferum*, *Appl. Environ. Microbiol.*, **34**, 282-285.

Flüher, H., Stolzy, L. H. and Ardakani, M. S. (1976) A statistical approach to define soil aeration in respect to denitrification, *Soil Sci.*, **122**, 115-123.

Focht, D. D. and Verstraete, W. (1977) Biochemical ecology of nitrification and denitrification, *Adv. Microb. Ecol.*, **1**, 135-214.

Franco, A. A. (1978) Contribution of the legume–*Rhizobium* symbiosis to the ecosystem and food production, in Döbereiner, J., Burris, R. H., Hollaender, A., Franco, A. A., Neyra, C. A. and Scott, D. B. (eds) *Limitations and Potentials for Biological Nitrogen Fixation in the Tropics*, New York, London, Plenum Press, 65-74.

French, J. R. J., Turner, G. L. and Bradbury, J. F. (1976) Nitrogen fixation by bacteria from the hindgut of termites, *J. Gen. Microbiol.*, **95**, 202-206.

Gottschalk, G. (1979) *Bacterial Metabolism*, New York, Heidelberg, Berlin, Springer-Verlag.

Greenwood, D. J. (1969) Effect of oxygen distribution in the soil on plant growth, in Whittington, W. J. (ed.) *Root Growth*, New York, Plenum Press, 202-223.

Hahn, J. and Junge, C. (1977) Atmospheric nitrous oxide: A critical review, *Z. Naturforsch.*, **32A**, 190-214.

Haines, B. L. (1977) Nitrogen uptake. Apparent pattern during old field succession in Southeastern U. S., *Oecol.*, **26**, 295-303.

Hall, J. B. (1978) Nitrate-reducing bacteria, in *Microbiology-1978*, Washington, D.C., American Society for Microbiology, 296-298.

Hasan, S. M. and Hall, J. B. (1975) The physiological function of nitrate reduction in *Clostridium perfringens*, *J. Gen. Microbiol.*, **87**, 120-128.

Heal, O. W. and MacLean, S. F. (1975) Comparative productivity in ecosystems – secondary productivity, in van Dobben, W. H. and Lowe-McConnell, R. H. (eds) *Unifying Concepts in Ecology*, The Hague. Dr. W. Junk Publishers, 89-108.

Helyar, K. R. (1976) Nitrogen cycling and soil acidification, *J. Austr. Inst. Agr. Sci.*, **42**, 217-221.

Hesselman, H. (1917) Om våra skogsföryngringsåtgärders inverkan på salpterbildningen i marken och dess betydelse för barrskogens föryngring, *Medd. Stat. Skogsforsk. anst.*, **13-14**, 923-1076.

Higinbotham, N. B. (1973) The mineral absorption process in plants, *Bot. Rev.*, **39**, 15-69.

Hynes, R. K. and Knowles, R. (1978) Inhibition by acetylene of ammonia oxidation in *Nitrosomonas europea*, *FEMS Letters*, **4**, 319-321.

Klingmüller, W. (1979) Genetic engineering for practical application, *Naturw.*, **66**, 182-189.

Kluyver, A. J. and Verhoeven, W. (1954) Studies on true dissimilatory nitrate reduction, II, The mechanism of denitrification, *Ant. van Leeuwenh. J. Microbiol. Serol.*, **20**, 241-262.

Knowles, R. (1978) Common intermediates of nitrification and denitrification and the metabolism of nitrous oxide, in *Microbiology-1978*, Washington, D.C., American Society for Microbiology, 367-371.

Lee, J. A. and Stewart, G. R. (1978) Ecological aspects of nitrogen assimilation, *Adv. Bot. Res.*, **6**, 1-43.

Lees, H. (1954) The biochemistry of the nitrifying bacteria, in *Autotrophic Microorganisms*, Symposium of the Society for General Microbiology, 4, 84-98.
Likens, G. E., Bormann, F. H., Pierce, R. S., Eaton, J. S. and Johnson, N. M. (1977) *Biogeochemistry of a Forested Ecosystem*, New York, Heidelberg, Berlin, Springer-Verlag.
Liu, S. C., Cicerone, R. J., Donahue, T. M. and Chameides, W. L. (1977) Sources and sinks of atmospheric N_2O and the possible ozone reduction due to industrial fixed nitrogen fertilizers, *Tellus*, 29, 251-263.
Losada, M. (1975/76) Metalloenzymes of the nitrate-reducing system, *J. Molecular Catalysis*, 1, 245-264.
Lübbers, D. W. (1969) The principle of construction and application of various Pt-electrodes, *Prog. Respir. Res.*, 3, 136-146.
McElroy, W. B., Elkins, J. W., Wofsy, S. C. and Yung, Y. L. (1976) Sources and sinks for atmospheric N_2O, *Rev. Geophys. Space Phys.*, 14(2), 143-150.
Miflin, B. J. and Lea, P. J. (1977) The pathway of nitrogen assimilation in plants, *Progr. Phytochem.*, 4, 1-26.
Moore, A. W. (1969) *Azolla:* Biology and agronomic significance, *Bot. Rev.*, 35, 17-34.
NAS (1978) *Nitrates: An Environmental Assessment*, Washington, D.C., National Academy of Sciences.
Newton, J. W. and Herman, A. I. (1979) Isolation of cyanobacteria from the aquatic fern *Azolla, Arch. Microbiol.*, 120, 161-165.
Neyra, C. A. and Van Berkum, P. (1977) Nitrate reduction and nitrogenase activity in *Spirillum lipoferum., Canad. J. Microbiol.*, 23, 306-310.
Nicholas, D. J. D. (1978) Intermediary metabolism of nitrifying bacteria, with particular reference to nitrogen, carbon, and sulfur compounds, in *Microbiology-1978*, Washington, D.C., American Society for Microbiology, 305-309.
Odum, E. P. (1969) The strategy of ecosystem development, *Science*, 164, 262-270.
Parnas, H. (1975) Model for decomposition of organic material by microorganisms, *Soil Biol. Biochem.*, 7, 161-169.
Parnas, H. (1976) A theoretical explanation of the priming effect based on microbial growth with two limiting substrates, *Soil Biol. Biochem.*, 8, 139-144.
Parsons, J. W. and Tinsley, J. (1975) Nitrogenous substances, in Gieseking, J. E. (ed.), *Soil Components*, 1, 263-304.
Payne, W. J., (1973) Reduction of nitrogenous oxides by microorganisms, *Bact. Rev.*, 37, 409-452.
Persson, T. (1980) Contribution of soil fauna to nitrogen mineralization in a coniterous forest. Possible effects of food selection by the fauna, in Rosswall, T. (ed.) *Processes in the Nitrogen Cycle*, SNV PM 1213, Stockholm, Swedish Environment Protection Board (in Swedish), 135-141.
Postgate, J. R. (1977) Consequences of the transfer of nitrogen fixation genes to new hosts, *Ambio*, 6, 178-180.
Purchase, B. S. (1974) The influence of phosphate defiency on nitrification, *Plant and Soil*, 41, 541-547.
Rice, E L. and Pancholy, S. K. (1972) Inhibition of nitrification by climax ecosystems, *Amer. J. Bot.*, 59, 1033-1040.
Rice, E. L. and Pancholy, S. K. (1973) Inhibition of nitrification by climax ecosystems, II, *Amer. J. Bot.*, 60, 691-702.
Ritchie, G. A. F. and Nicholas, D. J. D. (1972) Identification of the sources of nitrous oxide produced by oxidation and reductive processes in *Nitrosomonas europea, Biochem. J.*, 126, 1181-1191.

Robinson, E. and Robbins, R. C. (1970) Gaseous nitrogen compounds pollutants from urban and natural sources, *J. Air Poll. Contr. Ass.*, **20**, 303–306.

Rosswall, T. (1976) The internal nitrogen cycle between microorganisms, vegetation and soil, in Svensson, B. H. and Söderlund, R. (eds) Nitrogen phosphorus and sulphur–Global cycles, SCOPE Report 7, *Ecol. Bull. (Stockholm)*, **22**, 157–167.

Rosswall, T. (1978) Impact of massive microbe-mediated transformations on the global environment: Microbial activity affecting the thickness of the ozone layer and the CO_2 concentration in the atmosphere, paper presented at the XII International Congress of Microbiology, SCOPE/UNEP International Nitrogen Unit Report 5.

Rosswall, T. (1979a) Nitrogen losses from terrestrial ecosystems–global, regional and local considerations, in *Proc. V International Meeting on Global Impacts of Applied Microbiology*, Bangkok, 17–26.

Rosswall, T. (1979b) Applied microbiology can aid developing countries, *Ambio*, **8**, 116–117.

Rosswall, T. (1980) Aquatic microbial ecology–concepts and trends, in *Microbiology-1980*, Washington, D.C., American Society for Microbiology.

Rosswall, T. and Vitousek, P. M. (in press) Priorities in nitrogen cycling research, in Rosswall, T. (ed.) *Nitrogen Cycling in West African Ecosystems*, Proc. SCOPE/UNEP Workshop, Ibadan, Nigeria, December 1978, published by the Royal Swedish Academy of Sciences, Stockholm.

Rosswall, T., Flower-Ellis, J. G. K., Johansson, L. G., Jonsson, S., Rydén, B. E. and Sonesson, M. (1975) Stordalen (Abisko), Sweden, in Rosswall, T. and Heal, O. W. (eds) Structure and function of tundra ecosystems, *Ecol. Bull. (Stockholm)*, **20**, 265–294.

Rosswall, T., Lohm, U. and Sohlenius, B. (1977) Développement d'un microcosm pour l'étude de la mineralisation et de l'absorption radiculaire de l'azote dans l'humus d'une forêt de conifères (*Pinus sylvestris* L.), *Lejeunia N. S.*, **84**, 1-24.

Ruschel, A. P., Victoria, R. L., Salati, E. and Henis, Y. (1978) Nitrogen fixation in sugarcane (*Saccharum officinarum* L.), in Granhal, U. (ed.) Environmental role of nitrogen-fixing blue-green algae and asymbiotic bacteria, *Ecol. Bull. (Stockholm)*, **26**, 297–303.

Sanni, S. O. (1976) Vesicular-arbuscular mycorrhiza in some Nigerian soils and their effect on the growth of cowpea (*Vigna unguiculata*), tomato (*Lycopersicon esculentum*) and maize (*Zea mays*), *New Phytol.*, **77**, 667–671.

Söderlund, R. and Svensson, B. H. (1976) The global nitrogen cycle, in Svensson, B. H. and Söderlund, R. (eds) Nitrogen, phosphorus and sulphur–global cycles, SCOPE Report 7, *Ecol. Bull. (Stockholm)*, **22**, 23–73.

Sprent, J. I., Scott, R. and Perry, K. M. (1978) The nitrogen economy of *Myrica gale* in the field, *J. Ecol.*, **66**, 657–668.

Staaf, H. and Berg, B. (1977) Mobilization of plant nutrients in a Scots pine forest moor in central Sweden, *Silva Fennica*, **11**, 210–217.

Stewart, W. D. P. (1977) Present-day nitrogen-fixing plants, *Ambio*, **6**, 166–173.

Stevenson, F. J. (1972) Nitrogen: Element and geochemistry, in Fairbridge, R. W. (ed.) *The Encyclopedia of Geochemistry and Environmental Sciences*, New York, Cincinnati, Toronto, London, Melbourne, Van Nostrand Reinhold, Volume IV, 795–801.

Suzuki, I. (1978) Microbial inorganic oxidations with special reference to NH_3, paper presented at the XII International Congress of Microbiology, München, FRG, September 1978.

Sweeney, R. E., Liu, K. K. and Kaplan, I. R. (1978) Oceanic nitrogen isotopes and their uses in determining the source of sedimentary nitrogen, in Robinson, B. W. (ed.) *Stable Isotopes in the Earth Sciences,* DSIR Bulletin **220**, 9–26, Wellington, New Zealand Department of Scientific and Industrial Research.

Talley, S. N., Talley, B. and Rains, D. W. (1977) Nitrogen fixation by *Azolla* in rice fields, in Hollaender, A. (ed.) *Genetic Engineering for Nitrogen Fixation,* New York, London, Plenum Press, 259–281.

Ulrich, B. (1978) A sytems approach to the role of nutrients in controlling rehabilitation of terrestrial ecosystems, in *The Breakdown and Restoration of Ecosystems,* New York, London, Plenum Press, 105–121.

Vitousek, P. (1980) Nitrogen losses from disturbed ecosystems—Ecological considerations, in Rosswall, T. (ed.) *Nitrogen Cycling in West African Ecosystems,* Stockholm, Royal Swedish Academy of Sciences.

Vitousek, P. M., Gosz, J. R., Grier, C. C., Melillo, J. M. Reiners, W. A. and Todd, R. L. (1979) Nitrate losses from disturbed ecosytems, *Science,* **204**, 469–474.

Wilson, T. R. S. (1978) Evidence for denitrification in aerobic polagic sediments, *Nature,* **274**, 354–356.

Yatazawa, M. and Yoshida, S. (1979) Stem nodules in *Aeschynomene indica* and their capacity of nitrogen fixation, *Physiol. Plant,* **45**, 293–295.

Yoshida, T. and Alexander, M. (1970) Nitrous oxide formation by *Nitrosomonas europea* and heterotrophic microorganisms, *Soil Sci. Soc. Amer. Proc.,* **34**, 880–882.

Yoshinari, T. and Knowles, R. (1976) Acetylene inhibition of nitrous oxide reduction by denitrifying bacteria, *Biochem. Biophys. Res. Commun.,* **69**, 705–710.

Zablotowicz, R. M., Eskew, D. L. and Focht, D. D. (1978) Denitrification in *Rhizobium, Canad. J. Microbiol.,* **24**, 757–760.

Zavitkovski, J., Hansen, E. A. and McNeel, H. A. (1979) Nitrogen-fixing species in short rotation systems for fiber and energy production, in Gordon, J. C., Weeler, C. T. and Perry, D. A. (eds) *Symbiotic Nitrogen Fixation in the Management of Temperate Forests,* Corvallis, Oregon Forest Research Laboratory, Oregon State University, 388–402.

Zumft, W. G. and Cárdenas, J. (1979) The inorganic biochemistry of nitrogen bioenergetic processes, *Naturwissenschaften,* **66**, 81–88.

CHAPTER 3

Current Problems Related to the Atmospheric Part of the Sulphur Cycle

H. RODHE

Department of Meteorology, University of Stockholm, Sweden

ABSTRACT

The problem of estimating emissions into the atmosphere of reduced sulphur compounds such as hydrogen sulphide (H_2S) and dimethyl sulphide (DMS) is briefly discussed. Several recent measurements confirm the existence of such emissions but quantifications remain quite uncertain. Measurements of sulphur dioxide (SO_2) in the upper troposphere indicate the existence of a significant source of SO_2 at those levels (other than H_2S or DMS).

The most obvious and most well documented result of man's impact on the sulphur cycle is the existence of large areas around industrialized regions with increased concentrations of SO_2 and sulphate in air, and of acid in precipitation. These acid 'blotches' today occur at least in Europe and in the eastern parts of North America. If man-made emissions of SO_2 are permitted to increase further the future will hold larger and more such acid blotches with their detrimental effects on the environment.

3.1 INTRODUCTION

This paper contains a brief discussion about a few aspects of the atmospheric part of the sulphur cycle which I consider to be of particular importance, and where more information is needed for a proper evaluation of this cycle and, in particular, of man's impact on it. It is not intended as a review of the state of the art concerning sulphur in the atmosphere. For such reviews the reader is referred to the Proceedings of the International Symposium on Sulfur in the Atmosphere (ISSA, 1978). Chapter 4 in this volume, Ivanov (1981), contains a summary of the work carried out by a group of Soviet scientists as part of the SCOPE project 1.3, 'Global biochemical sulphur cycle', which was initiated after the publication of SCOPE 7 (1976).

As in the previously mentioned publications the emphasis will be on regional and global scales. The special problems associated with urban areas with high concentrations of SO_2 due to human activities will not be discussed.

3.2 EMISSIONS OF REDUCED SULPHUR COMPOUNDS

One of the most important uncertainties in quantitative estimates of the sulphur cycle has been and still is the emission of H_2S and other volatile reduced sulphur compounds from soils, vegetation, and waters into the atmosphere. Such emissions may be the result of decomposition of organic sulphur compunds originating from 'assimilatory sulphate reduction', of sulphate reduction in connection with decomposition of organic matter under anaerobic conditions ('dissimilatory sulphate reduction') or of volcanic emissions.

Whereas earlier estimates of such fluxes were obtained essentially by indirect means (cf. Eriksson, 1963, as an example) a few more direct estimates have been made in recent years. Such estimates have been based on flux chamber measurements of H_2S emissions from the surface (Hansen et al., 1978), measurements of profiles of dimethyl sulphide (DMS) in ocean surface waters (Nguyen et al., 1978), and measurements of H_2S in the atmosphere (Jaeschke et al., 1978, Slatt et al., 1978). These measurements unambiguously show that emissions of reduced sulphur compounds do take place. However, the problem of making quantitative estimates of these fluxes, particularly as aggregates over extended areas and over longer time periods, is not an easy one.

For example, Hansen et al. (1978) reported emissions of H_2S from two tide pools in Denmark of 0.06 to 1.6 mmol m^{-2} h^{-1} (daily averages for summer conditions). The sites were specifically selected for expected high values of the emission. The question immediately arises of the areal extent of such conditions. Can such measurements at a few sites be used to estimate fluxes over larger regions and maybe even over the whole globe? Evidently the uncertainties in such estimates are bound to be very large. How do the total amounts from such high emission areas of limited extent compare with emissions of lower density but covering much larger areas? I think that this type of question needs to be more thoroughly discussed before regional and global emission rates can be reliably estimated from flux chamber measurements.

Estimates of emissions from the ocean surface may, of course, be more readily extrapolated to large areas because of the larger horizontal homogeneity of the ocean. It should be possible to take differences due to changes in temperature and biological productivity into account.

Measurements of H_2S in the atmosphere may be used to make estimates of emission rates either by the gradient method (Jaeschke et al., 1978) or by treating the atmosphere (or part of it) as a box and using the relation $F = M/\tau$ where M is the total mass of H_2S inside the box and τ the average residence of time of H_2S (Slatt et al., 1978). The former technique requires horizontal homogeneity and steady-state conditions. In view of areal variability of emissions the first condition may be difficult to fulfil, particularly over land. The latter requires an integration over a large horizontal area and over the whole depth of the atmosphere, plus an estimate of the residence time.

Because of these difficulties, more measurements of H_2S (and other gaseous reduced sulphur compounds) in the atmosphere are a necessary requirement for a better understanding of the role of biogenic sulphur emissions. In particular, vertical profiles up to a height of several kilometres and over different types of surface should be given high priority.

In this connection, a plea for better estimates of man-made emissions of H_2S should be made. Such emissions most likely represent a negligible contribution to regional or global sulphur budgets but they may give rise to locally significant concentrations. An identification of these sources is necessary for a proper interpretation of the relation between measured concentrations of H_2S and natural sources (cf. Jaeschke et al., 1978).

Graedel (1979) recently made an attempt to put together data on sulphur compounds in the marine atmosphere. Using information about H_2S concentrations in the air (Slatt et al., 1978) and DMS concentrations in air (Maroulis and Bandy, 1978) and in surface waters (Nguyen et al., 1978) in a photochemical model of the marine atmosphere he showed that fluxes of H_2S and DMS from the surface of about Tg-S yr^{-1} of each of these two compounds are consistent with present knowledge about their oxidation rates in the air and with estimated concentrations of the oxidants (mainly OH). However, the estimate by Östlund and Alexander (1963) of a life time of H_2S in surface water of the ocean of less than one hour is difficult to reconcile with the idea of a substantial emission of H_2S from the ocean surface. Also, the SO_2 concentrations predicted by Graedel's model were much lower than those measured by, for example, Maroulis et al. (in press). It is evident that our knowledge about sulphur in the marine atmosphere is far from satisfactory.

Excess sulphate (i.e. sulphate that does not originate from sea salt) in air and in precipitation in locations very remote from industrial regions may also be used to estimate the importance of natural sulphur emissions into the atmosphere. Figure 3.1, taken from Delmas (1979), shows a compilation of sulphate data from Greenland ice samples. Since local sources may be excluded, the existence of significant concentrations a few hundred years ago provides a strong indication that natural sulphur sources (other than sea spray) do not make up a substantial part of global budgets. On the other hand, the increase during recent years is consistent with the growing importance of man-made emissions. Variations in the frequency of occurrence of volcanic emissions should, however, also be taken into account in the interpretation of such data. It is possible that the levels after 1963 could have been influenced by the great volcanic eruption of Mount Agung (Delmas, 1979). More historical data of the kind shown in Figure 3.1 may provide very valuable information in relation to the atmospheric part of the sulphur cycle.

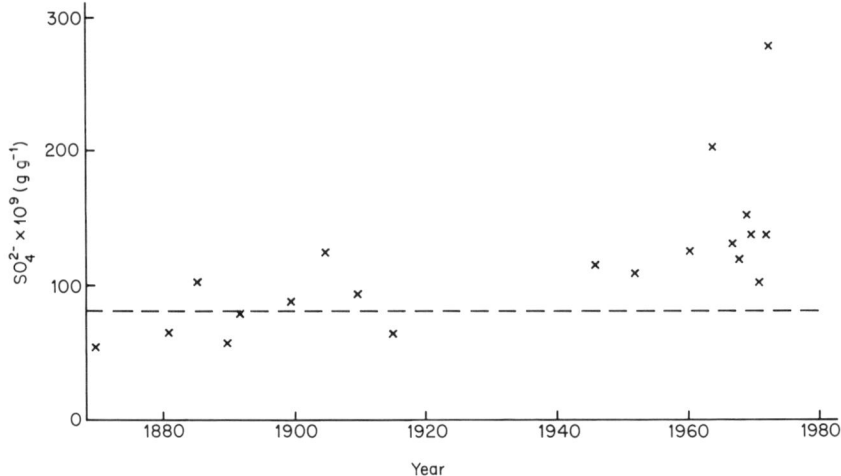

Figure 3.1 Data on sulphate from Greenland snow and ice samples compiled by Delmas (1979)

3.3 SULPHUR FLUX INTO THE UPPER TROPOSPHERE

The flux of sulphur compounds into the upper troposphere and the stratosphere is of particular interest in connection with the possible impact of sulphate particles on climate. At these heights the scavenging of sulphate particles by atmospheric precipitation is probably slow enough for the sulphate to become reasonably well mixed zonally around the globe. To the extent that the distribution is not too variable in the latitudinal direction, it here becomes meaningful to talk about global, or at least hemispheric, fluxes and concentrations. This is one good reason—maybe the only one?—for establishing global sulphur budgets.

Unfortunately, very few measurements of sulphur compounds have been made at heights between 5 and 10 km. The most comprehensive set of measurements of SO_2 was reported recently by Maroulis *et al.* (in press).

The circled numbers in Figure 3.2 show some results of their measurements obtained during flights in the boundary layer and at a height of 5-6 km mainly over the Pacific Ocean in connection with the GAMETAG experiment. High values in mid-latitudes of the northern hemisphere most likely reflect man-made emissions in the industrialized regions of North America, Europe, and Japan.

The relatively high concentrations of SO_2 (50-100 pptv) occurring at 5-6 km in tropical latitudes and in the southern hemisphere raise a difficult problem of interpretation. It is hardly possible that SO_2 at those latitudes could originate from the industrialized regions in mid-latitudes. Since the characteristic time for mixing across the equator is at least several months almost all SO_2 would have been transformed to sulphate (and probably removed by precipitation) long before entering the

Current Problems Related to the Atmospheric Part of the Sulphur Cycle 55

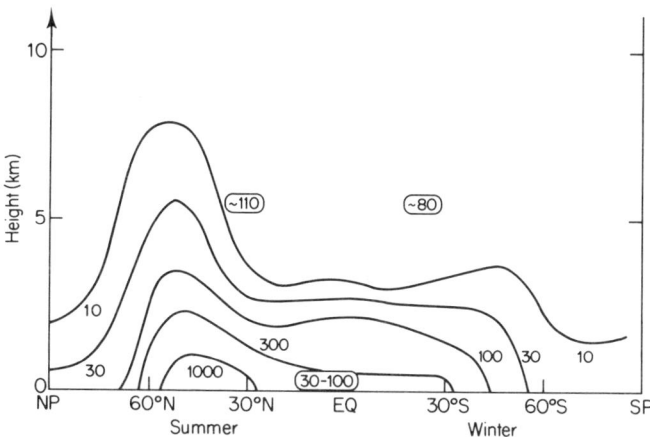

Figure 3.2 Concentrations of SO_2 in units of pptv: observed (Maroulis *et al.*, 1980), encircled numbers; estimated in a photochemical transport model (Rodhe and Isaksen, in press), isolines

southern hemisphere. If the source was located on the ground, within the same latitude belt, one would expect high concentrations to occur at lower elevations. However, the measurements by Maroulis *et al.* (in press) in the lowest kilometre over the Pacific Ocean give values that are in fact slightly lower than those obtained at 5-6 km.

Even if the SO_2 is formed *in situ* from a precursor which is emitted at the ground one would still expect a decrease of its concentration with height, at least as long as the precursor is shortlived enough to exhibit a pronounced decrease in concentration with height. In view of the short residence time of DMS and H_2S—of the order of a day—it is hardly possible that these gases can make up for the observed vertical distribution of SO_2. In fact, calculations in a global two dimensional transport model carried out by Rodhe and Isaksen (in press) predict SO_2 concentrations at 5 km in the tropics of at most a few pptv. Figure 3.2 shows the distribution of SO_2 in July estimated in their model. It is based on an assumed emission of 80 Tg SO_2-S from man-made sources and 40 Tg H_2 S-S from natural sources.

Carbonyl sulphide, OCS, is a sulphur compound which is relatively inert in the troposphere but which is broken down by ultra-violet radiation in the lower stratosphere. It thereby provides a high altitude source for SO_2. This SO_2 is thought to be a contributing factor in the formation of sulphate particles at around 20 km (Crutzen, 1976). A certain production of SO_2 also takes place in the troposphere through the reaction of OCS with the OH radical. However, it seems very unlikely that SO_2 produced this way could explain concentrations of about 100 pptv in the middle troposphere (Rodhe and Isaksen, in press).

3.4 SULPHUR IN INDUSTRIALIZED REGIONS

Annual average pH of precipitation in northwest Europe during 1974 is shown in Figure 3.3. A similar pattern occurs over the eastern US and southeastern Canada (see chapter 6, this volume). Although solid information is lacking about the preindustrial situation, there are strong indications that these acid regions are to a high degree caused by man-made emissions of sulphur and nitrogen oxides. The dominant anions in such acid precipitation are sulphate and nitrate.

In Europe the ratio of the average molar concentration of sulphate to that of nitrate is about 1.5 and on an equivalent basis the ratio is thus about 3 (Söderlund, 1977). In the eastern US the relative importance of sulphate is somewhat lower (Likens et al., 1977).

Judging from observations of the chemical composition of precipitation in northern Europe, there has been a general increase in the concentration of sulphate and nitrate over the last 25 years. The positive trend has been more pronounced for nitrate than for sulphate so that the sulphate/nitrate ratio has been going down (Rodhe et al., 1981). These trends are largely consistent with estimated trends in rates of man-made emissions of sulphur and nitrogen oxides (OECD, 1977). However, the question has been raised (Granat, 1978) as to why the observed trends in sulphate concentration—as well as in deposition—during the sixties and early seventies at many places in northern Europe did not follow the upward trend in SO_2 emissions as reported in the OECD study on long range transport of air pollutants (Fjeld, 1976). Actually sulphate concentrations seem to have remained approximately constant at many stations and even to have declined slightly at some. This apparent contradiction has not yet been satisfactorily resolved. In considering this problem the following factors should be kept in mind.

(i) The emission trends reported by Fjeld (1976) represent an aggregate over the whole of Europe. In order for a comparison with observed data to be relevant one would have to make estimates of how the total deposition over the region has varied. No such attempt has been made so far. To interpret data from individual stations one has to be more specific about the emissions in those areas that affect the particular stations. For example, the emission trends reported by Fjeld (1976) are evidently not at all representative for the United Kingdom where emissions seem to have declined since the early sixties (Reed, 1978). It is evident that for this kind of trend evaluation, emission data must be divided into subregions. Such an attempt has recently been done for the eastern US by Husar et al. (1979).

(ii) Year-to-year variations in meteorological parameters such as wind directions and precipitation amounts may significantly affect even annual deposition values. For example, Munn and Rodhe (1971) showed that annual values of sulphate deposition in southern Sweden were positively correlated with the frequency of occurrence of rain bearing storms with trajectories from the main source areas in UK and at the European continent.

(iii) Emissions by man of other compounds than sulphur may have had an effect

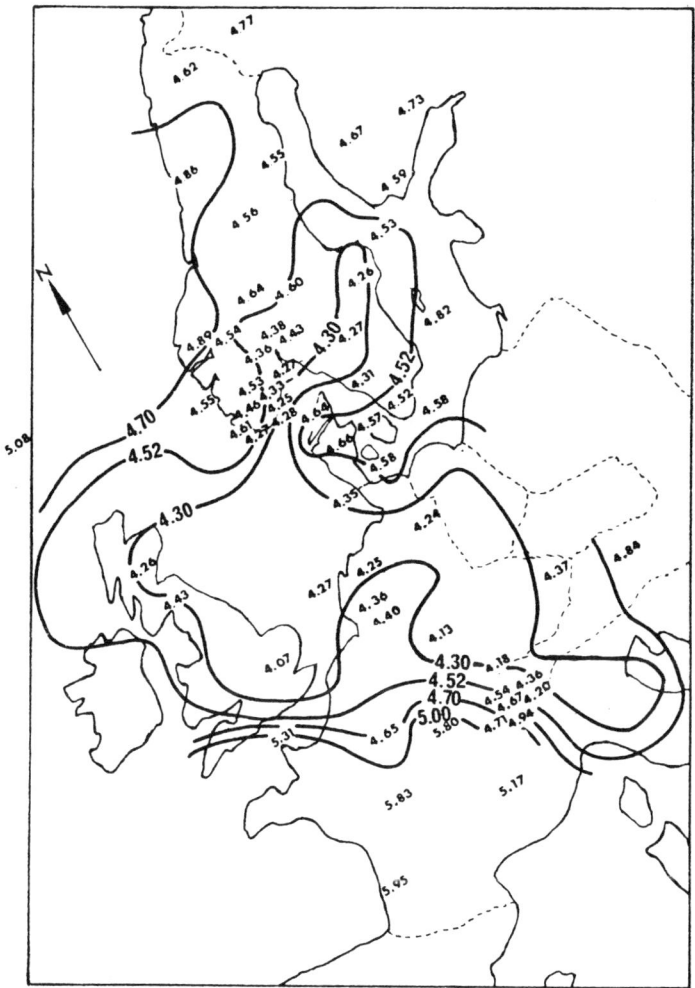

Figure 3.3 Annual average pH of precipitation in Europe for the period July 1972 to March 1975 (OECD, 1977)

on the relation between emission and deposition trends of this compound. This is because a changing chemical climate significantly affects transformation and removal rates and thereby the source-receptor relationship. For example, a lowering of the pH of cloud droplets will have a tendency to slow down the aqueous phase oxidation of SO_2 to sulphate (Penkett *et al.*, 1979). Similarly, increasing emissions of NO_x may reduce the atmospheric concentration of the OH radical and thereby slow down the gas phase oxidation of SO_2 (cf. Rodhe *et al.*, 1981).

3.5 INTERACTIONS WITH OTHER CYCLES

It is notable that as far as gas phase chemistry is concerned, sulphur compounds do not seem to have any appreciable impact on the other cycles. If the particle and liquid phase are taken into account the situation is different. For example, a large fraction of atmospheric ammonia is probably incorporated into sulphate aerosols to form NH_4HSO_4 and $(NH_4)_2SO_4$. An increased amount of sulphate in the atmosphere may therefore directly affect parts of the nitrogen cycle by *reducing* the atmospheric residence time of gaseous ammonia. Since aerosol particles of submicron size generally have a longer residence time in the atmosphere than reactive gases such as NH_3, the formation of ammonium sulphates at the same time *increases* the total residence time of ammonium compounds (Söderlund, 1977).

As indicated at the end of section 3.4, the cycling of sulphur through the atmosphere may be significantly affected by changes in the concentrations of other compounds. In their study of formation of sulphuric acid and nitric acid during long range transport, Rodhe *et al.* (1981) demonstrated that the competition between SO_2 and NO_x for the available OH radicals constitutes an important indirect link between the sulphur and nitrogen cycles in the atmosphere on local and regional scales. If emissions of NO_x are increased in an industrial region, the consumption of OH radicals at the transformation of NO_2 to HNO_3 tends to reduce the concentration of OH in the atmosphere. This, in turn, will slow down the gas phase transformation of SO_2 to H_2SO_4, thereby reducing the concentration of H_2SO_4 at least during the first few tens of hours of travel time.

Another coupling between the atmospheric parts of the sulphur and nitrogen cycles is the possible revolatilization of HNO_3 from aerosols made acid by sulphuric acid. An observed anticorrelation between sulphate and nitrate in atmospheric aerosols is an indication of such a process.

These examples point to the need for considering sulphur not as a passive component in atmospheric chemistry but as a truly interactive part. Another coupling which needs to be much better understood and which may turn out to be of great significance is the influence of deposition of sulphuric acid on soil and sediment processes. For example, it has been observed that an acidification of terrestrial ecosystems leads to an increased mobility of metals (Tyler, 1978) with subsequent risks for toxic concentrations in biota. Also, the generation of key species such as N_2O and CH_4 may be or become affected at least within the most heavily acidified regions of the world (Bolin and Arrhenius, 1977). As these regions grow in number and in size this may potentially become a global concern.

3.6 OUTLOOK FOR THE FUTURE

It is clear that man is stongly influencing the atmospheric part of the sulphur cycle in certain regions of the world and that this influence has a dramatic effect on the chemical climate in those regions. In many other parts of the world the sulphur cycle is very much closer to its natural state.

It has been a common exercise to sum the various fluxes into global totals and to take such values to represent some kind of average conditions. The problem is that such average values are representative *neither* of the heavily polluted regions *nor* of the cleaner parts of the world. A more realistic global perspective of the sulphur cycle is to look at the globe as at least moderately healthy in large areas but with a few serious acid blotches at present covering a few per cent of the area of the globe.

A natural question to pose is: What is going to happen if man-made sulphur emissions are permitted to increase substantially in the future? Evidently there will be more acid blotches forming as a result of industrialization in other parts of the world. For reasons discussed in the previous sections the present blotches of acid precipitation will probably not be much more intensive—i.e. will not achieve a lower pH in the precipitation—but will instead grow in size. However, the dry deposition of SO_2, which accounts for a substantial portion of the total deposition of acidifying compounds, will probably increase roughly in proportion to the emissions of SO_2.

It is important to realize that a development like the one painted above is by no means inevitable. Even if the consumption of fossil fuels continues to grow for a considerable time, corresponding increases in the emissions of sulphur compounds may well be avoided by the application of known industrial purification techniques. Up to now strong incentives have largely been lacking for combatting the acidification problem rationally. It is our duty as scientists to contribute to a better understanding of the dangers associated with human intervention in the sulphur cycle and thereby to lessen the likelihood that they will materialize.

To be able to provide solid data on changes in the chemical composition of air and of precipitation that might occur in industrialized regions it is absolutely essential to establish and continue long term measurement programmes. Such programmes are important not only in already industrialized regions but also in those areas where industrialization is likely to take place in the future and particularly where the resilience to acid deposition may be low.

3.7 REFERENCES

Bolin, B., and Arrhenius, E. (eds) (1977) Nitrogen—An essential life factor and a growing environmental hazard, *Ambio*, **6**, 96–105.

Crutzen, P. J. (1976) The possible importance of CSO for the sulfate layer of the stratosphere, *Geophys. Res. Lett.*, **3**, 73–76.

Delmas, R. (1979) Sulphate in polar snow and ice, *Proc. International Symposium on Sulphur Emissions and the Environment*, London, 8–10 May 1979, The Society of Chemical Industries, 72–76.

Eriksson, E. (1963) The yearly circulation of sulfur in nature, *J. Geophys. Res.*, **68**, 4001–4008.

Fjeld, B. (1976) Forbruk av fossilt brensel i Europa og utslipp av SO_2 i perioden 1900–1972 (Consumption of fossil fuels in Europe and emissions of SO_2 during the period 1900–1972), Norwegian Institute for Air Research, Teknisk Motat 1/76.

Graedel, T. E. (1979) Reduced sulfur emissions from the open oceans, *Geophys. Res. Lett.,* **6**, 329–331.
Granat, L. (1978) Sulfate in precipitation as observed by the European Atmospheric Chemistry Network, *Atmospheric Environment,* **12**, 413–424.
Hansen, M. H., Ingvorsen, K., and Jørgensen, B. B. (1978) Mechanisms of hydrogen sulfide release from coastal marine sediments to the atmosphere, *Limnol. Oceanogr.,* **23**, 68–76.
Husar, R. B., Patterson, D. E., Holloway, J. M., Wilson, W. E., and Ellestad, T. G. (1979) Trends of eastern US haziness since 1948, in *Proc. Fourth Symposium on Atmospheric Turbulence, Diffusion and Air Quality,* Reno, Nevada, January 1979.
ISSA (1978) *Sulfur in the Atmosphere: Proceedings of the International Symposium,* Dubrovnik, Yugoslavia, 7–14 September, 1977 (Husar, R. B., Lodge, J. P. Jr., and Moore, D. J. eds), Oxford, Pergamon Press.
Jaeschke, W., Georgii, H. -W., Claude, H., and Malewski, H. (1978) Contributions of H_2S to the atmospheric sulfur cycle, *Pageoph.,* **116**, 465–475.
Likens, G. E., Bormann, F. H., Pierce, R. S., Eaton, J. S., and Johnson, N. M. (1977) *Biogeochemistry of a Forested Ecosystem,* New York, Springer-Verlag.
Maroulis, P. J., and Bandy, A. R. (1977) Estimate of the contribution of biologically produced dimethyl sulfide to the global sulfur cycle, *Science,* **196**, 647–648.
Maroulis, P. J., Torres, A. L., Goldberg, A. B., and Bandy, A. R. (1980) Measurements of tropospheric background levels of SO_2 on Project GAMETAG, *J. Geophys. Res.,* in press.
Munn, R. E., and Rodhe, H. (1971) On the meteorological interpretation of the chemical composition of monthly precipitation samples, *Tellus,* **23**, 1–13.
Nguyen, B. C., Gandry, A., Bonsang, B., and Lambert, G. (1978) Reevaluation of the role of dimethyl sulphide in the sulphur budget, *Nature,* **275**, 637–639.
OECD (1977) *The OECD Programme on Long Range Transport of Air Pollutants: Measurements and Findings,* Paris, Organization for Economic Cooperation and Development.
Östlund, H. G., and Alexander, J. (1963) Oxidation rate of sulfide in sea water, a preliminary study, *J. Geophys. Res.,* **68**, 3995–3997.
Penkett, S. A., Jones, B. M. R., Brice, K. A., and Eggleton, A. E. J. (1979) The importance of atmospheric ozone and hydrogen peroxide in oxidizing sulphur dioxide in cloud and rainwater, *Atmospheric Environment,* **13**, 123–137.
Reed, L. E. (1978) SO_2 emissions in the UK, *Nature,* **273**, 334.
Rodhe, H., Crutzen, P. J., and Vanderpol, A. (1981) Formation of sulfuric and nitric acid in the atmosphere during long range transport, *Tellus,* **33**, in press.
Rodhe, H., and Isaksen, I. (1980) Global distribution of sulfur compounds in the troposphere estimated in a height/latitude transport model, *J. Geophys. Res.,* in press.
SCOPE (1976) Nitrogen, phosphorus and sulphur—Global cycles, in Svensson, B. H., and Söderlund, R. (eds) SCOPE Report 7, *Ecol. Bull. (Stockholm),* **22**.
Slatt, B. J., Natusch, D. F. S., Prospero, J. M., and Savoie, D. L. (1978) Hydrogen sulfide in the atmosphere of the Northern Equatorial Atlantic Ocean and its relation to the global sulfur cycle, *Atmospheric Environment,* **12**, 981–991.
Söderlund, R. (1977) NO_x pollutants and ammonia emissions—A mass balance for the atmosphere over NW Europe, *Ambio,* **6**, 118–122.
Tyler, G. (1978) Leaching rates of heavy metal ions in forest soil, *Water, Air, and Soil Pollution,* **9**, 137–148.

Some Perspectives of the Major Biogeochemical Cycles
Edited by Gene E. Likens
© 1981 SCOPE

CHAPTER 4

The Global Biogeochemical Sulphur Cycle*

M. V. IVANOV

*Institute of Biochemistry and Physiology of Microorganisms,
USSR Academy of Sciences, Pushchino, USSR*

ABSTRACT

The major sulphur fluxes of both natural and anthropogenic origin have been assessed by generalization of the results of our studies and data available in the literature. Annually about 120 Tg S are extracted by man from the lithosphere in fossil fuels and sulphur-containing raw materials for the chemical industry. Of this amount 70 Tg S are emitted to the atmosphere with the products of fuel combustion. About half of the remaining 50 Tg S directly enters rivers with sewage and residual waters, and another part is applied with fertilizers to agricultural land. Parallel with anthropogenic sulphur, volcanic gases contribute markedly to the atmospheric sulphur cycle over continents where, according to our data, the sulphur flux amounts to 29 Tg yr^{-1}. The major transfer of sulphur from continents to the ocean is brought about by river runoff with an annual sulphur load of 224 Tg in which the sulphur of anthropogenic pollution accounts for 109 Tg. Tremendous amounts of sulphur are involved in processes of internal turnover between the oceanic atmosphere and its waters. The total flux of various sulphur forms (organic, sulphate, and pyrite) from oceanic water to sediments and further to the lithosphere amounts to 130 Tg yr^{-1}; Thus, our estimates point to the fact that the anthropogenic sulphur fluxes to the atmosphere and hydrosphere have reached a level comparable with that of natural fluxes. The analysis of prospects for future uses of various fossil fuels and fertilizers suggests that by the end of this century the anthropogenic sulphur fluxes will increase notably in all regions of the world.

The present report is a preliminary account of the work of Soviet scientists on preparing the SCOPE report as part of project 1.3, 'The global biogeochemical sulphur cycle'. The resolution on conducting this work was adopted at the Paris meeting of the SCOPE Executive Committee in May 1977 and supported by the Open Session of the SCOPE Executive Committee in London (October 1977).

*The work on this project was carried out with the assistance of Prof. V. A. Grinenko and Drs. A. A. Migdisov (Institute of Geochemistry and Analytical Chemistry), A. Yu. Lein (Institute of Biochemistry and Physiology of Microorganisms), A. G. Ryaboshapko (Institute of Applied Geophysics), I. I. Volkov and A. G. Rozanov (Institute of Oceanology), and A. L. Rabinovich (Hydrochemistry Institute).

Table 4.1 Estimates of Annual Sulphur Fluxes according to Various Authors (Tg yr^{-1}, 1978)

Source of suphur	Eriksson (1960, 1963)	Robinson and Robbins (1968, 1970)	Kellogg et al. (1972)	Friend (1973)	Granat et al. (1976)	Our data
Biological decay (land)	110	68		58	5	23
Biological decay (ocean)	170	30	90	48	27	19
Volcanic activity	–	–	1.5	2	3	29
Sea spray (total)	45	44	47	44	44	60
Anthropogenic emission	40	70	50	65	65	70
Precipitation (land)	65	70	86	86	43	
Dry deposition (land)	100	20	10	20		103
Absorption (vegetation)	75	26	15	15	28	
Precipitation and dry deposition (ocean)	100	71	72	71	73	90
Absorption (ocean)	100	25	–	25		
Total sulphur involved in atmospheric balance	365	212	183	217	144	
Land–sea transfer	–10	+26	+ 5	+ 8	+18	+21
Sea–land transfer	5	4	4	4	17	6
Fertilizers	10	11	–	26	–	24
Rock weathering	15	14	–	42	–	61
Total river runoff	80	73	–	136	122	224

4.1. THE PRESENT STATE OF THE PROBLEM AND OBJECTIVES

At the time our work started, several surveys had already been published assessing the content of sulphur compounds in the atmosphere, hydrosphere, and lithosphere, and considering magnitudes of some sulphur fluxes (Eriksson, 1963; Friend, 1973; Kellogg et al., 1972; Robinson and Robbins, 1970; Granat et al., 1976). Particular attention was paid to the sulphur cycle in the atmosphere; the remaining fluxes were considered only to balance the atmospheric cycle (see Table 4.1). Therefore, many of the fluxes were calculated arithmetically, from values obtained by analysis of the processes occurring in the atmosphere. Because of the relatively numerous and informative data on the atmospheric processes as compared to the less investigated processes occurring in the lithosphere and hydrosphere, such an approach is not only reasonable but also the only one possible. However, this approach leads to considerable variations in estimates of the other biogeochemical processes of the sulphur cycle, as can be seen clearly from the data of Table 4.1.

If we consider that the estimate of the atmospheric sulphur cycle is based on rather limited data with a number of *a priori* assumptions, there arises a natural desire, or rather a vital necessity, to find an independent approach to the estimation of the separate fluxes which together constitute the entire global biogeochemical sulphur cycle.

Our first task was to compile the primary information on the content of different sulphur forms in separate reservoirs and geospheres and to estimate the magnitudes of each flux, where possible, by different methods. These several approaches were based on the analytical geochemical data of the sulphur compounds and on the *in situ* experimental studies which constitute the basis of dynamic biogeochemistry.

In recent years the ever-growing interest in global cycles is explained not only by our rapidly increasing scientific knowledge in this field but also by the eagerness of the scientists to investigate the quantitative and global processes involved in the cycles of major elements. Such interest is stimulated by fears that the scope of industrial and agricultural activity of man may have an unpredictable impact on the environment not only in the industrialized regions of the earth but also on a global scale.

It is noteworthy that the possible consequences of global pollution of the environment by sulphur compounds have not attracted as much attention as, for example, the investigations of atmospheric pollution by anthropogenic CO_2. Both scientific and popular literature widely discuss the possible climatic and ecological impacts of CO_2 accumulation in the atmosphere. But pollution of the atmosphere and hydrosphere by such chemically active compounds as hydrogen sulphide and sulphur oxides presents even now a real problem for vast regions of industrialized countries. Many ecological aspects of this problem have been treated in detail by Nriagu (1978b, 1978c).

For example, the Scandinavian countries and some other industrialized countries of Europe and North America are now facing a problem of pollution of river and lake waters and soil by fallout of sulphuric acid.

It is fairly simple to estimate the total anthropogenic sulphur flux on a regional and global scale. For this purpose it is sufficient to use information on the amount of the different kinds of sulphur-containing raw materials consumed by different branches of industry and agriculture.

Such data, particularly the anthropogenic sulphur flux to the atmosphere, are available in many published studies (see Table 4.1). The variation in the estimates is more conditioned by the use of data from different years than by the methods of approach of the various authors. In other words, these differences reflect an objective increase in consumption rate of sulphurous raw materials and, primarily, of fossil fuel over the last two decades.

The quantitative estimate of different natural processes involved in the global sulphur cycle appears much more complicated. For many of these estimates there is not only a lack of factual material, but also an absence of methodical approach to the estimate. Therefore, we sought, where possible, not only to indicate the natural and anthropogenic contributions to the sulphur fluxes separately, but also paid particular attention to the natural processes.

Finally, one more feature of our work was an attempt to estimate the fluxes of some other elements involved with the biogeochemistry of the sulphur compounds. As seen from Figure 4.1 all of the main reactions of the sulphur cycle involving living organisms are closely related to the carbon cycle.

The amount of carbon involved in the fluxes of the sulphur cycle through biogenic processes varies depending on the type of organisms undertaking the metabolism of the sulphur compounds. In the processes of bacterial chemosynthesis, which are characterized by low amounts of energy utilized for the CO_2 assimilation, only relatively small amounts of carbon are transformed into organic matter.

In anaerobic, bacterial photo-assimilation of CO_2 where sulphur compounds are used as electron donors, the amounts of oxidized sulphur and assimilated carbon are comparable. In anaerobic sulphate-reduction, 24 g of organic carbon are mineralized for each 32 g of reduced sulphate sulphur. Thus, in ecosystems with an advanced development of photoautotrophic and sulphate-reducing bacteria both groups of microorganisms transform significant amounts of carbon compounds and, consequently, these organisms should be considered not only as participants in the sulphur cycle but as active biogeochemical agents of the carbon cycle.

The extremely important geochemical role of plant photosynthesis in both carbon and sulphur cycles is evident [see Figure 4.1 (VIII) and chapter 6, this volume], although sulphur is not reduced in photosynthesis and sulphur-organic compounds are formed in secondary metabolic processes.

The role of key reactions in the sulphur cycle on the geochemistry of different metals is important and diverse. Hydrogen sulphide released in sulphate reduction plays a principal role in the immobilization of metals as sulphides, most of which are insoluble in water. On the other hand, sulphuric acid and dissolved metals are formed in the weathered zone by biological and chemical oxidation of metal sulphides and elemental sulphur. During intensive weathering by acid sulphate, e.g. in volcanic sulphur-sulphide deposits, sulphuric acid decomposes not only the ores but also the volcanic rocks (Lein and Ivanov, 1970; Ivanov, 1971). As a result the sulphate solutions are enriched in Al^{3+}, Fe^{2+}, Fe^{3+}, Ti^{2+} and even in silicic acid.

The great importance of many reactions of the sulphur cycle for the turnover of oxygen is indisputable. In many aquatic reservoirs much of the dissolved oxygen is used to oxidise hydrogen sulphide and elemental sulphur, and these reactions influence significantly the oxygen cycle and overall geochemistry in reservoirs (Kuznetsov, 1970).

Thus, our attention has been focused on the less investigated problems of the global sulphur cycle: the direct estimate of fluxes, the separate estimate of natural

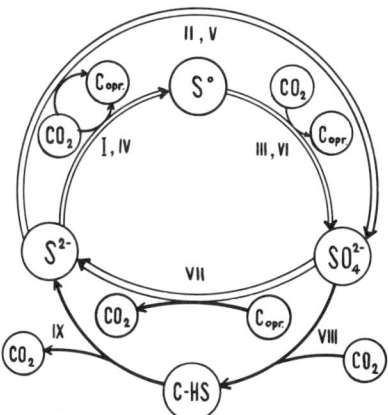

Figure 4.1 Relationship between general biological processes of the sulphur cycle and some reactions of the carbon cycle.

Chemoautotrophic bacteria

$$2H_2S + O_2 = 2S^0 + 2H_2O \quad (I)$$
$$H_2S + 2O_2 = SO_4^{2-} + 2H^+ \quad (II)$$
$$2S^0 + 3O_2 + 2H_2O = 2SO_4^{2-} + 2H^+ \quad (III)$$

Photoautotrophic bacteria

$$2H_2S + CO_2 = 2S^0 + (CH_2O) + H_2O \quad (IV)$$
$$H_2S + 2CO_2 + H_2O = SO_4^{2-} + (CH_2O) + 2H^+ \quad (V)$$
$$S^0 + 2CO_2 + 2H_2O = SO_4^{2-} + 2(CH_2O) + 2H^+ \quad (VI)$$

Sulphate-reducing bacteria

$$SO_4^{2-} + 2C_{org} = S^{2-} + 2CO_2 \quad (VII)$$

Photosynthesis (VIII)
Putrefaction (IX)

and anthropogenic components of individual fluxes and the relation of the sulphur cycle to the cycles of other elements.

4.2 GLOBAL SULPHUR CYCLE, ANTHROPOGENIC FLUXES

A summary diagram of the global sulphur cycle with quantitative estimates of the sulphur fluxes is given in Figure 4.2. The numbers near the arrows designate the

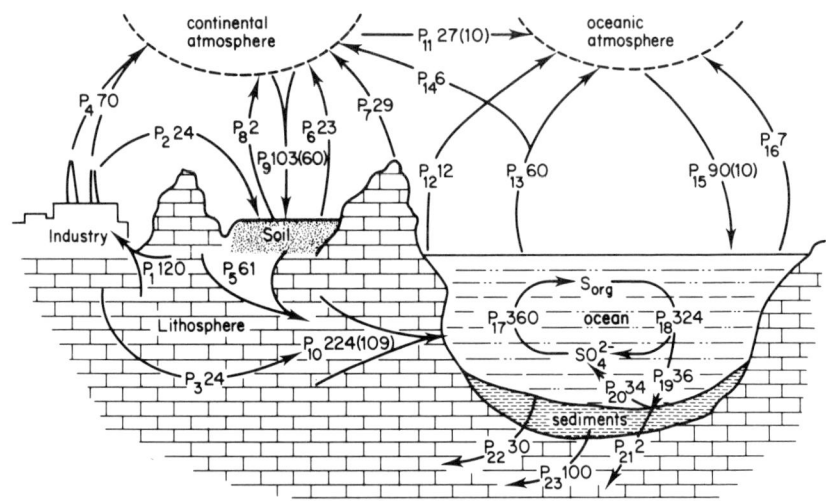

Figure 4.2 Fluxes of the global biogeochemical sulphur cycle. The figures following the flux numbers designate the magnitude of a total sulphur flux, while those in parentheses show anthropogenic contributions.

Symbols: P_1—sulphur flux from the lithosphere due to all kinds of mining; P_2—sulphur flux to the soil with fertilizers; P_3—sulphur flux with industrial sewage waters; P_4—flux of anthropogenic sulphur to the atmosphere; P_5—sulphur flux from water erosion processes; P_6—flux of biogenic sulphur; P_7—flux of volcanic sulphur; P_8—sulphur of dust emission; P_9—sulphur flux to land with atmospheric precipitation; P_{10}—sulphur flux with river runoff; P_{11}—anthropogenic and natural sulphur flux from continents to ocean; P_{12}—flux of biogenic H_2S from coastal shallow sediments; P_{13}—flux of marine sulphur with the sea-spray; P_{14}—flux of marine sulphur to continents; P_{15}—sulphur flux from the oceanic atmosphere to ocean; P_{16}—emission of reduced sulphur from the oceanic surface; P_{17}—sulphur in biomass of marine plants; P_{18}—mineralized sulphur of dead marine plants and other organisms; P_{19}—flux of organic sulphur to the sea bottom; P_{20}—organic sulphur oxidized to sulphate and returned to sea water; P_{21}—organic sulphur buried in marine sediments; P_{22}—sulphate sulphur buried in marine sediments; P_{23}—reduced sulphur buried in marine sediments

total sulphur flux in Tg S yr^{-1} for all compounds. The contributions from anthropogenic activities are indicated by numbers in parentheses. So the flux P_1 is the sum value of sulphur extraction from all kinds of mining on the globe. The main part of this flux (about 70 Tg S yr^{-1}) is from the fossil fuel and polymetallic sulphide ores. The combustion of fuel and metal smelting from ores are accompanied by an intensive emission of sulphur oxides to the atmosphere (flux P_4).

We consider that all sulphur of the flux P_4 enters the continental atmosphere and mixes therein with other sulphur forms from the natural sources (see fluxes P_6, P_7, P_8, P_{14}). Most of the sulphur of the continental atmosphere returns as dry and wet deposition on to the continental surface (flux P_9), and a smaller part enters the oceanic atmosphere (flux P_{11}).

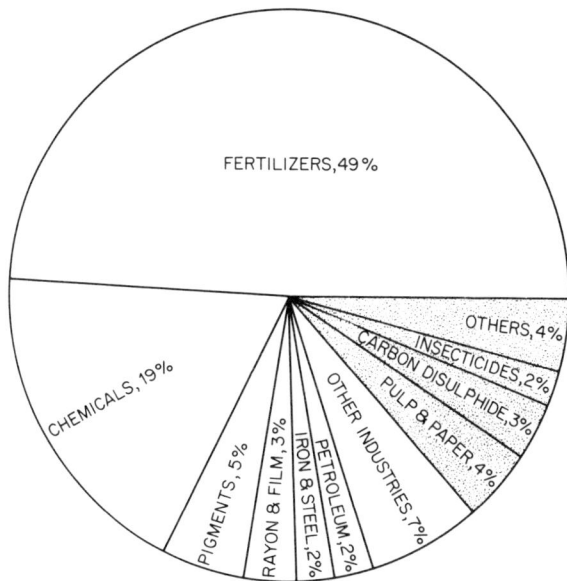

Figure 4.3 Uses of sulphur during 1970 (after Cote, 1970). The hatched area represents uses in the form of sulphuric acid

Two other anthropogenic fluxes—P_2 and P_3—are the final by-product of processing the sulphurous raw materials in different branches of industry. The sulphurous raw materials (some 50 Tg yr^{-1}) are supplied by the mining industry in the form of pyrites, elemental sulphur, or sulphur from gas deposits (Nriagu, 1978a).

According to Cote's data (1970) cited in the work of Nriagu (1978a) about half of the sulphur extracted in the world in 1970 was used for production of fertilizers and (see Figure 4.3), consequently, brought with them in to the soil (flux P_2). Then this sulphur appeared in the water draining from the soil and joined the river runoff (P_{10}). The second half of the sulphur obtained from pyrites, sulphate, and gas deposits is consumed by the chemical industry and finally is discharged with domestic or industrial sewage (flux P_3) and also enters the rivers.

Two important fluxes of anthropogenic sulphur through the soil (P_2) and sewage (P_3) entering the runoff are added with another 60 Tg of anthropogenic sulphur falling on to the land surface with precipitation (see the flux P_9). So, currently rivers annually bring about 108 Tg of anthropogenic sulphur to the ocean.

Moreover, 10 Tg of anthropogenic sulphur, as a minimum, is transferred by air flows from the continental atmosphere to the atmosphere over the ocean (flux P_{11}). This sulphur also finally enters the ocean waters. Thus, the annual anthropogenic pollution of the ocean by sulphur compounds amounts to 120 Tg S yr^{-1} and exceeds the magnitude of natural river runoff of sulphur compounds (see section 3).

Table 4.2 Average Content and Weight Ratio of Sulphates/Chlorides in the River Waters of the Various Basins of the USSR (according to Alyokin and Brajnikova, 1964)

Basin	Content (mg litre^{-1})		SO_4^{2-}/Cl^-
	SO_4^{2-}	Cl^-	
Siberian rivers			
The Chukotsk, East Siberian,	17.5	17.7	0.99
Kara and Laptev Seas	8.8	7.3	1.20
Rivers of the European part of the USSR			
The Baltic, Barents,	7.0	4.0	1.75
White, Black, Azov and	14.8	5.0	2.96
Caspian Seas	41.5	16.5	2.51
	62.1	18.9	3.28

4.3 NATURAL SULPHUR IN RUNOFF

The estimate of the natural sulphur flux by rivers (Figure 4.2, P_{10}) is based on the well-known data of Livingstone (1963) and the results given by Alyokin and Brajnikova (1964) on the ionic runoff of the USSR rivers. Since these data characterize both the runoff from the regions polluted by anthropogenic sulphur and from non-polluted areas, to estimate the natural sulphur flux it was necessary to find an objective index for the low level of water pollution. When analyzing the data relating to the USSR's rivers (Table 4.2) we noticed that in the runoff of Siberian rivers the weight ratio of sulphates/chlorides was close to unity, whereas for the European rivers of the USSR where industry was more developed in the catchment, the relative sulphate content was much higher.

If we use the same approach for analysing Livingstone's data for the different continents, in the rivers of Africa, Asia, and South America with relatively low water pollution the ratio of sulphates/chlorides is also close to unity, whereas in the rivers of North America and, particularly, in those of Europe the waters are considerably enriched in sulphate (Table 4.3).

Having assumed that the weight content of sulphates is equal to chlorides in the river waters of all continents and using the recent data about volume of river runoff taken from the monograph 'World Water Budget and Water Resources of the Earth' (Korzun et al., 1974), we obtained a value for the yearly runoff of natural sulphur of 104 Tg (Table 4.4).

The separate sulphur fluxes of the runoff both natural and anthropogenic are shown in Figure 4.4. We would only draw attention to the grand total: 224 Tg S yr^{-1}, of which 109 Tg is anthropogenic sulphur.

Table 4.3 Average Content and Weight Ratio of Sulphates/Chlorides in River Waters of Various Continents (according to Livingstone, 1963)

Continent	Content (mg litre^{-1})		SO_4^{2-}/Cl^-
	SO_4^{2-}	Cl^-	
Europe	24.0	6.9	3.48
North America	20.0	8.0	2.50
South America	4.8	4.9	0.98
Asia	8.4	8.7	0.96
Africa	13.5	12.1	1.11
Australia	2.6	10.0	0.26

Table 4.4 Sulphur Flux to the Ocean in the Form of Dissolved Sulphates in River Runoff

Continent	Area having runoff to the ocean (millions of km^2)[a]	Annual runoff (thousands of km^3)[a]	Average SO_4^{2-} content (mg litre^{-1})	Annual runoff (Tg)	
				sulphate	sulphur
N. America	19.5	7.84	8.0[b]	62.72	20.90
S. America	16.4	11.70	4.3[c]	50.31	16.77
Euro-Asia	39.1	16.40	8.4[d]	137.76	45.92
Africa	20.5	4.11	13.5[e]	55.48	18.49
Australia and Oceania	4.8	2.37	2.6[e]	6.16	2.05
Total	100.3	42.42	–	312.43	104.13

[a]According to Korzun et al., 1974;
[b]estimated from chlorides (Table 4.3);
[c]according to Gibbs, 1972;
[d]Livingstone's data for Asia;
[e]Livingstone, 1963.

4.4 ATMOSPHERIC SULPHUR CYCLE

In the upper and right-hand parts of Figure 4.2 are given some principal fluxes of the atmospheric sulphur cycle. We shall consider only the fluxes which differ considerably from those published previously. First of all for flux P_7 the emission of volcanic sulphur, our estimate exceeds by a factor of 10 and even more the data published by Kellogg et al. (1972), Friend (1973), and Granat et al. (1976). Such a discrepancy is not spurious. The above-cited authors took into account only the emissions of hydrogen sulphide and sulphur oxides occurring during volcanic eruptions. In contrast, we also considered the quantitative estimates of all year-round

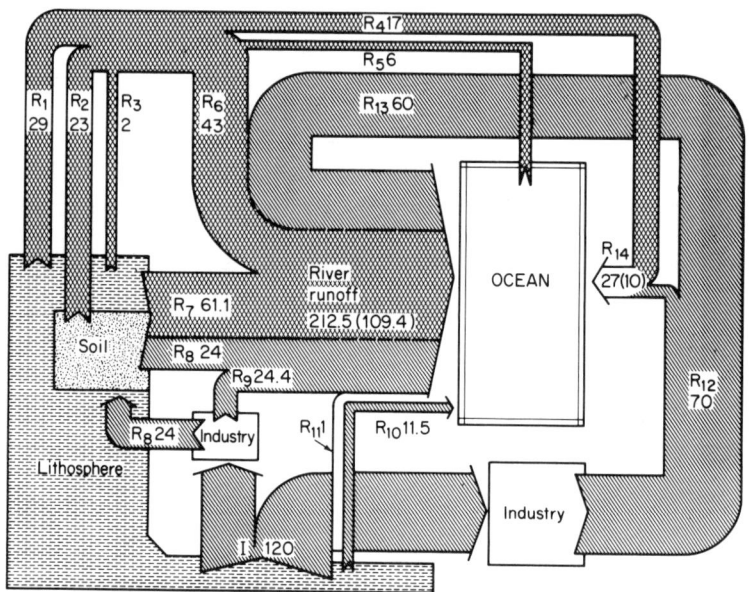

Figure 4.4 Scheme showing the major natural and anthropogenic sulphur fluxes contributing to the flux of sulphate sulphur from continents to oceans. Figures by the flux indices denote the amount of sulphur in Tg yr^{-1}. Symbols: R_1–R_7, natural sulphur fluxes: R_1, volcanic sulphur; R_2, biogenic sulphur; R_3, sulphur of dust emission; R_4, sulphur flux from continental atmosphere to oceanic; R_5, marine sulphur flux to continent; R_6, total flux of natural sulphur from atmosphere to continents; R_7, sulphur of water erosion processes; R_8, sulphur transferred to soil drainage waters with fertilizers; R_9, sulphur of sewage waters; R_{10}, sulphur of underground flux; R_{11}, sulphur of sewage waters of mining manufacturing; R_{12}, anthropogenic sulphur flux to atmosphere; R_{13}, atmospheric sulphur precipitating to continents; R_{14}, anthropogenic and natural sulphur transfer from continents to ocean.

I, sulphur output from lithosphere as a result of mining; II, processing of pyrites, native sulphur and sulphur-containing gases in chemical industry; III, fossil fuel combustion and sulphide ore processing

sulphur emissions from the fumaroles, hot springs, and lakes made by a number of Soviet, American, and Japanese investigators (Berlyand, 1975; Okita and Shimozuru, 1975; Stoiber and Jepsen, 1973).

To estimate the emission of biologic sulphur to the atmosphere was quite difficult and was complicated by the variation in estimated values from different surveys, that is from 170 Tg S yr^{-1} (Eriksson, 1960, 1963) to 27 Tg S yr^{-1} (Granat et al., 1976). Throughout this range the estimates were not obtained from factual data but by difference from the atmospheric sulphur budget.

The factual data from the measurement of biogenic emissions of hydrogen sulphide from the shallow sediments of Limfiord were obtained four years ago by the Danish ecologists Hansen et al. (1978) and Jørgensen (1977, 1978). In parallel with these experiments Jørgensen obtained a vast amount of factual material on the intensity of sulphate reduction in the sediments of Limfiord using radioactive sulphate. By comparing the intensity of sulphate reduction in shallow coastal samples from various parts of the ocean with direct measurements of hydrogen sulphide emission to the atmosphere we managed to estimate this flux (Figure 4.1, P_{12}). The maximum value for this estimate would be 10-12 Tg S yr^{-1} (Ivanov, 1979). To confirm this value we need, however, to obtain some additional experimental data on the emission of hydrogen sulphide to the atmosphere in various climatic zones.

Two more fluxes of biogenic sulphur (Figure 4.2, Flux P_6 from land to atmosphere, 23 Tg; and flux P_{16} from sea to atmosphere, 7 Tg) were calculated from the actual content of reduced sulphur in the atmosphere. The analysis of results obtained by Brazilian hydrochemists (Brinkmann and Santos, 1974) on the hydrogen sulphide emission from the flood plain of the Amazon River to the atmosphere shows that the sulphur flux from the flood-plain surface of 6000 km^2 0.2-0.25 Tg during only four months (Ivanov, 1979). The other estimates of atmospheric sulphur fluxes obtained in our version of the global sulphur cycle approximate more or less to the values cited in the review of Swedish investigators published by SCOPE in 1976 (Granat et al., 1976).

4.5 SULPHUR CYCLE IN THE OCEAN

Previous works on the global sulphur cycle have usually considered only the fluxes connecting oceanic and atmospheric sulphur reservoirs (Figure 4.2, P_{13}, P_{15}, P_{16}). The internal oceanic sulphur cycle, as far as we know, was not assessed quantitatively. However, this cycle includes important processes such as the incorporation of sulphur into biomass of sea plants (Figure 4.2, P_{17}) and the reverse process of mineralization of organic sulphur compounds after aerobic and anaerobic decay of dead plants and other organisms (Figure 4.2, P_{18}).

This part of the global sulphur cycle was evaluated from data on the carbon cycle in the ocean. The calculations were based on results published by Romankevich (1977). From these data the annual primary production of organic matter in the ocean is 36 000 Tg. Let us assume that the sulphur content of phytoplankton biomass is one per cent, then about 360 Tg of sulphate sulphur of sea water is accumulated by living biomass annually (Figure 4.2, P_{17}).

After death the decay of organisms proceeds mainly in oceanic water. Not more than 10 per cent of the organic matter synthesized by phytoplankton is assumed to reach the bottom as particles (Romankevich, 1977). From this analysis, flux P_{19} is 36 Tg of sulphur per year, and flux P_{18} reaches 324 Tg per year.

Organic sulphur compounds which arrive at the bottom of the ocean are intensively decayed by microorganisms in the uppermost horizons of the sediments.

Table 4.5 Total Sulphate Sulphur Flux from Water to the Oceanic Sediments

Forms of sulphate in sediments	Mass of sulphur (Tg yr^{-1})
In solid phase of clay sediments	11.9
In silt water	7.1
In form of barium sulphate (baryte)	4.0
In biogenic carbonates	1.4

About 90-95 per cent of the organic sulphur is oxidized there to sulphate which returns to the water (flux P_{20}). The rest of the sulphur, about 2 Tg, is buried with silt in the sediments as organic sulphur compounds (flux P_{21}).

Of special interest for estimation of the global sulphur cycle are the data on sulphate and sulphide sulphur fluxes from the water column of the ocean to the sediments (Figure 4.2, fluxes P_{22} and P_{23}). During the present period of geological history the formation of evaporites occurs less extensively, primarily in some continental basins (Strakhov, 1960). Nevertheless, sulphate sulphur is buried with the sediments in the oceans as skeletons, shells and remains of various organisms and as minerals of terrigenic drifts. In addition some sulphates are buried in dissolved form in the interstitial waters of sediments. The total value of this sulphate flux (P_{22}), calculated by Dr. I. Volkov specially for this report, is 30 Tg per year. The components of this flux are tabulated in Table 4.5.

The world oceanological and geochemical literature contains several important works on the biogeochemical processes of sulphate reduction in seas and oceans, and the distribution of reduced forms of sulphur in modern sediments (Strakhov, 1960, 1972; Berner, 1971; Goldhaber and Kaplan, 1974). However, in none of these papers was the global sulphur cycle evaluated quantitatively. Only a few attempts to calculate the annual formation of hydrogen sulphide have been made, for example in the Black Sea (Datsko, 1959; Skopintsev, 1975).

In both these works the annual production of hydrogen sulphide in the sea was estimated at 2 Tg by Datsko (1959) and 3.2 Tg by Skopintsev (1975). Datsko's caclulations are based on the balance of organic matter in the Black Sea, while Skopintsev used experimental data on the intensity of sulphate reduction published by Sorokin (1962).

These workers considered the processes of hydrogen sulphide formation in the whole basin including the water column and sediments; a possible burial of a part of the hydrogen sulphide and its derivatives in sediments was not taken into account. Therefore from these data it is impossible to evaluate how much sulphur leaves the cycle and is buried in newly formed sediments.

We can use several different techniques for quantitative evaluation of the reduced sulphur flux to bottom sediments.

Table 4.6 Amount of Reduced Sulphur Buried Annually in Sediments of the Baltic Sea (after Lein, 1978)

Method of flux calculation	Flux value (Tg yr^{-1})
1. Method of absolute masses: are of aleuritopelitic sediments 298 × 10^3 km^2, rate of dry sediment accumulation 10.87 g cm^{-2} per 1000 years, S content/H$_2$S 1.37% by dry weight of sediment	0.440
2. Method of annual inputs of material to the sea: total 46.10^6 Tg, S content/H$_2$S 1.37% by dry weight of sediment	0.580
3. Balance of sulphur fluxes in the runoff, precipitation from atmosphere, and water exchange through Danish channels	0.702
Average	0.574

1. Generalization of analytical data on the content of reduced sulphur in modern sediments of various types in different geomorphological areas of some seas and the world ocean (shelf, continental slope, etc.). This value, the rate of sediment formation, and the area of sedimentation allowed us to calculate the rate of annual sulphur flux to the sediments.

2. For more or less separated basins (inland seas, large gulfs of the ocean, etc.) this value may be obtained by multiplying the average sulphur content in reduced sediments by the amount of sedimentary material entering the basin from runoff and other sources.

3. The sulphur budget of some studied seas was calculated on the basis of the sulphate sulphur influx with surface and underground runoff and the amounts of reduced sulphur withdrawn from the sulphur cycle and buried in silts.

4. For a comparative analysis of the rate of sulphate reduction in various geomorphological zones and sediments of the ocean, the results on the sulphate reduction activity with radioactive sulphate were used (Ivanov, 1956, 1968, 1979).

5. To calculate the budget of sulphur compounds, in a number of cases the data on the balances of other components, such as organic carbon and barium were used.

To illustrate, some initial values and final results of the sulphur budget calculated for the Baltic sea are tabulated in Table 4.6. As seen from Table 4.6, the calculations done by these three different methods agree very closely.

It is noteworthy that the total flux of reduced sulphur alone to the sediments makes up 100 Tg per year (Figure 4.2, flux P_{23}). From the sum of fluxes P_{22} and P_{23}, about 130 Tg of sulphur in oxidized and reduced forms is transferred annually from the ocean to the lithosphere.

Therefore, these data make it possible to close the global biogeochemical sulphur cycle and to show that disregarding anthropogenic activity, the natural sulphur flux from the lithosphere, its main reservoir, is compensated by the reverse flux of sulphur compounds to the lithospheric sediments of the ocean.

4.6 QUANTITATIVE EVALUATION OF OTHER CYCLES FROM THE SULPHUR CYCLE

To illustrate possibilities for the quantitative evaluation of fluxes of other elements based on the sulphur fluxes let us consider only one example: the geochemical result of sulphate reduction in bottom sediments of the ocean. It is well known that pyrite is the main form of sulphur buried in sediments (Goldhaber and Kaplan, 1974; Volkov *et al.*, 1972, 1976).

Knowing the value of sulphur flux (Figure 4.2, P_{23}) we may assert that about 85 Tg of iron are preserved annually in sediments as pyrite.

From the same data and based on the sulphate reduction reaction (Figure 4.1, VII) it is possible to calculate the minimum quantity of organic carbon that would be mineralized during microbial sulphate reduction. It should be noted here, however, that the entire amount of H_2S formed during sulphate reduction in sediments would surely not be preserved as water insoluble compounds. Thus, to evaluate the amount of carbon involved in metabolism by sulphate reducing bacteria, it would be more correct to use values for the intensity of sulphate reduction. According to our calculations the annual value for sulphate reduction in the ocean sediments approximates 400 Tg. Therefore about 300 Tg of organic carbon is consumed during this process, i.e. 15-30 per cent of the total carbon deposited annually in the sediments of the ocean (Romankevich, 1977). Surely, such a large value is extremely important in the budget of organic matter or carbon in the ocean, as well as in the global sulphur cycle.

Carbon dioxide formed during mineralization of organic carbon in anaerobic conditions is responsible for the increase in the total alkalinity of interstitial waters in reduced sediments. One half of this carbon dioxide and bicarbonates enters the water, the other half is bound to calcium ions in water in the sediments. This partitioning results in the formation of diagenetic carbonates which have an anomalously light isotopic composition for carbon (Ivanov, 1968; Lein, 1978). As shown in the work of Lein (1978) in the Pacific and Indian oceans, the amount of such diagenetic carbonate in terrigenic sediments may be as high as 60 per cent of the total carbonate minerals found in sediments.

Therefore, a combined and systematic consideration of processes of the sulphur cycle and their connection with other elements gives valuable quantitative information on separate aspects of cycles for a number of elements including carbon, calcium, oxygen, and various metals.

4.7 CHANGES IN THE GLOBAL BIOGEOCHEMICAL SULPHUR CYCLE AS A RESULT OF MAN-MADE CONTRIBUTIONS

We have also studied the literature on possible changes in anthropogenic fluxes of sulphur during decades to come. The curves showing the increased output of raw

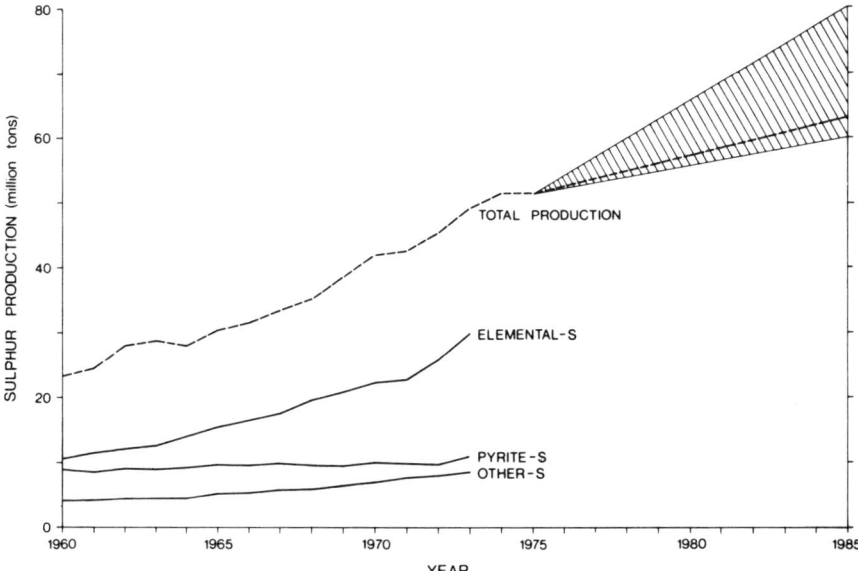

Figure 4.5 Production of sulphur from various sources since 1960. The projected worldwide production of sulphur in the coming decade is expected to lie within the hatched area; the historical trend line is shown as the broken curve (after Nriagu, 1978a)

materials containing sulphur for the chemical industry are given in Figure 4.5. As seen from the figure, the output of elemental, pyrite, and gaseous sulphur and production of sulphuric acid will increase by 30 per cent from 1975 to 1985. Nriagu (1978a) underlines also, that with agricultural development the consumption of sulphur will be greatly increased in developing countries of Africa and South America, where it will be double during 10 years. Thus, sulphur contamination of drainage waters and rivers will increase and will occur in new regions of the globe.

The curves in Figure 4.6 show the increase in atmospheric contamination with various energy sources. During conditions of an energy crisis and ever-rising prices of liquid and gaseous fuels the most realistic prognosis should take into account a substantial increase in coal consumption. As seen from the data presented, even with the present systems of fuel gas purification of SO_2, there is a most gloomy prospect for sulphur contamination of the global environment.

Actually, the correlation between values of natural and anthropogenic sulphur fluxes (Figure 4.2) shows that already by the mid-1970s the total amount of sulphur entering the atmosphere and hydrosphere has doubled since 1900 because of anthropogenous sulphur. Thus the natural biogeochemical systems for neutralization of sulphur oxides in the atmosphere and sulphuric acid in the hydrosphere and soils must already double their rate as compared to 1900.

The irregularity of the distribution of sulphur contamination source and the short average residence time of sulphur compounds, especially in the atmosphere,

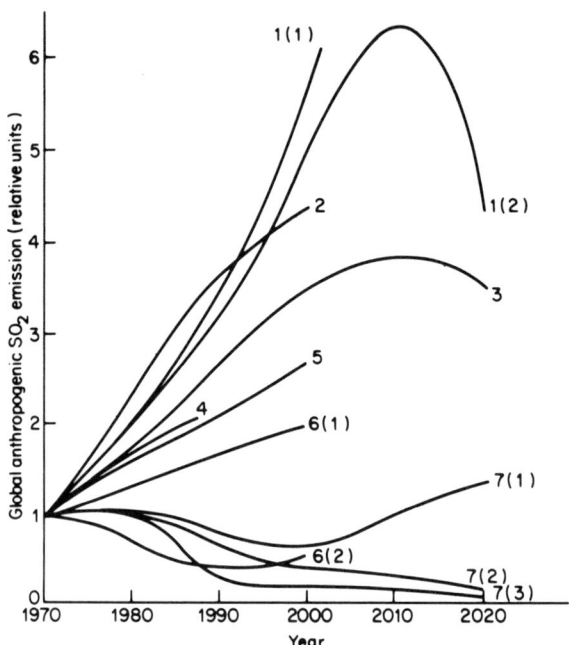

Figure 4.6 Predicted trends of global anthropogenic SO_2 emission into the atmosphere (after A. Ryaboshapko, 1978). 1, Spaite and Harrington, 1971 (209): 1(1), fossil fuel version: 1(2), fossil fuel–nuclear version with use of breeder-reactors; 2, Shvedov, 1976 (51), on basis of coal and oil use for energy; 3, this paper; 4, Roderick, 1975 (181); 5, Kellogg et al., 1972 (146), 6, Land, 1971 (150): 6(1), fossil fuel version without treating systems; 6(2), using treating systems; 7, Smill, 1975 (190): 7(1), fossil fuel version with treating systems; 7(2), fossil fuel–nuclear version; 7(3), nuclear version

enhance the menace of regional situations of crisis from sulphuric acid contamination of the atmosphere, hydrosphere, and soil.

The acidification of lake water and soil, the increase in sulphate content, and the appearance of hydrogen sulphide in river, lake, and even sea waters show that in a number of cases the natural systems are unable to manage the ever-increasing sulphur contamination.

A possible way out of the present situation may be the development and introduction of utterly new systems of full gas purification from sulphur oxides. The additional amount of sulphuric acid obtained may in turn decrease the demands for the expansion of pyrite and native sulphur outputs. As an alternative to sulphur contamination we may suggest the wide-spread development of nuclear power.

4.8 REFERENCES

Alyokin, O. A., and Brajnikova, L. U. (1964) *Runoff of Dissolved Compounds from the Territory of the USSR*, Moscow, Nauka (in Russian).
Berlyand, M. E. (1975) *Current Problems of Atmospheric Diffusion and Air Pollution*, Leningrad, Hydrometeoizdat (in Russian).
Berner, R. A. (1971) *Principles of Chemical Sedimentology*, New York, McGraw-Hill.
Brinkmann, W. L. E., and Santos, U. (1974) The emission of biogenic hydrogen sulfide from Amazonian floodplain lakes, *Tellus*, **26**(1-2), 262-267.
Cote, (1970) cited in Nriagu, 1978.
Datsko. V. G. (1959) *Organic Matter in the Southern Seas of the USSR*, Moscow, Izdat. Akad. Nauk. SSSR, (in Russian).
Eriksson, E. (1960) The yearly circulation of chloride and sulfur in nature: meteorological, geochemical and pedological implications, part 2, *Tellus*, **12**, 63-109.
Eriksson, E. (1963) The yearly circulation of sulfur in nature, *J. Geophys. Res.*, **68**(13), 4001-4008.
Friend, J. P. (1973) in *Chemistry of the Lower Atmosphere*, New York, London, Plenum Press.
Gibbs, R. J. (1972) Water chemistry of the Amazon River, *Geochim. Cosmochim. Acta*, **39**(9), 1061-1066.
Goldhaber, M. B., and Kaplan, I. R. (1974) The sulfur cycle, in *The Sea*, New York, Wiley, volume 5, 569-655.
Granat, L., Rodhe, H., and Hallberg, R. O. (1976) The global sulphur cycle, *Ecological Bull. (Stockholm)*, **22**, 89-134.
Hansen, M. H., Ingvorsen, K., and Jørgensen, B. B. (1978) Mechanisms of hydrogen sulfide release from coastal marine sediments to the atmosphere, *Limnol. Oceanogr.*, **23**(1), 68-76.
Ivanov, M. V. (1971) Bacterial processes in the oxidation and leaching of sulfide-sulfur ores of volcanic origin, *Chem. Geol.*, **7**, 185-211.
Ivanov, M. V. (1968) Microbial processes in the formation of sulphur deposits, *Israel Progr. Sci. Trans.*
Ivanov, M. V. (1979) Role of microorganisms in hydrogen sulfide formation, in *Role of Microorganisms in the Natural Cycles of Gases*, Moscow, Nauka, 114-130 (in Russian).
Ivanov, M. V. (1956) Use of isotopes for studying the sulfate-reduction process in Lake Belovod, *Mikrobiologia*, **25**(3), 305-309 (in Russian).
Jørgensen, B. B. (1978) A comparison of methods for the quantification of bacterial sulfate reduction in coastal marine sediments, I-III, *Geomicrobiol. J.*, **1**(1), 11-64.
Jørgensen, B. B. (1977) The sulfur cycle of a coastal marine sediment (Limfjorden, Denmark), *Limnol. Oceanogr.*, **22**(5), 814-832.
Kellogg, W. W., Cadle, R. D., and Allen, E. F. (1972) The sulphur cycle, *Science*, **175**, 587-596.
Korzun, A. M. *et al.* (1974) see: *World Water Budget and Water Resources of the Earth* (in Russian).
Kuznetsov, S. I. (1970) *Microflora of Lakes and Their Geochemical Activity*, Leningrad, Nauka (in Russian).
Land, G. W. (1971) Fossil fuel: national energy supply and air pollution, Pap. ASME, No. WA/FU-4, 1-15.
Lein, A. Yu. (1978) Formation of carbonate and sulfide minerals during diagenesis of reduced sediments, in *Environmental Biogeochemistry and Geomicrobiology*, Ann Arbor, volume 1, 339-354.

Lein, A. Yu. and Ivanov, M. V. (1970) Oxidation of elemental sulfur in vulcanogenic deposits of Kurilo-Kamchatsky region, in *Geology and Mineralogy of the Weathered Crust,* Moscow, Nauka, 182-213 (in Russian).

Livingstone, D. A. (1963) *Data of Geochemistry,* Sixth Edn., Chapter G., Chemical Composition of Rivers and Lakes, Geological Survey Prof. Pap. 440-G., Washington.

Nriagu, J. O. (1978a) Production and uses of sulfur, in *Sulfur in the Environment, part 1,* New York, Wiley, 1-21.

Nriagu, J. O. (ed.) (1978b) *Sulfur in the Environment, part 1, The Atmospheric Cycle,* New York, Wiley.

Nriagu J. O. (ed.) (1978c) *Sulfur in the Environment, part 2, Ecological Impact,* New York, Wiley.

Okita, T., and Shimozuru, D. (1975) Remote sensing measurements of mass flow of sulfur dioxide gas from volcanoes, *Bull. Volcanol. Soc. Jap.,* **19**(3), 151-157.

Roderick, H. (1975) Projected emission of sulphur oxides from fuel combustion in the OECD area 1972-1985, IPIECA Symposium, Tehran.

Robinson, E., and Robbins, R. C. (1968) *Emissions Concentrations and Fate of Gaseous Atmospheric Pollutants,* Menlo Park, California, Stanford Research Institute.

Robinson, E., and Robbins, R. C. (1970) Gaseous sulfur pollutants from urban and natural sources, *J. Air Pollut. Control. Assoc.,* **20**(4), 233-235.

Romankevich, E. A. (1977) *Geochemistry of Organic Matter in the Ocean,* Moscow, Nauka, (in Russian).

Shvedov, V. P. (1976) Progress of energetics and pollution of the biosphere, in *International Conference on Physical Aspects of Atmospheric Pollution* (1-20 June, 1974), Abstracts, Vilnus, 5-10 (in Russian).

Skopintsev, B. A. (1975) *Formation of Contemporary Chemical Composition of the Black Sea Water,* Leningrad, Hydrometeoizdat, (in Russian).

Smill, V. (1975) Energy and air pollution: USA 1970-2020, *J. Air Pollut. Control. Assoc.,* **25**(3), 233-236.

Sorokin, Yu. I. (1962) Experimental study of the bacterial reduction of sulfates in the Black Sea with the help of S^{35}, *Mikrobiologia,* **31**(3), 402-410 (in Russian).

Spaite, P. W. and Harrington, R. E. (1971) Abatement goes global, *Power Engineering,* **75**(2), 42-45.

Stoiber, R. E. and Jepsen, A. (1973) Sulfur dioxide contributions to the atmosphere by volcanoes, *Science,* **182**, 577-578.

Strakhov, N. M. (1960) *Bases of Lithogenesis,* Moscow, Izdat. Akad. Nauk. SSSR (in Russian).

Strakhov, N. M. (1972) Balance of reductive processes in sediments of the Pacific Ocean, *Litologia i Poleznye Iskopaemye,* **4**, 65-92 (in Russian).

Volkov, I. I., Rozanov, A. G., Zhabina, N. N. and Fomina, L. S. (1976) Sulfur compounds in sediments of the Californian Gulf and neighbouring parts of the Pacific, in *Biogeochemistry of Sediment Diagenesis in the Ocean,* Moscow, Nauka, 136-170 (in Russian).

Volkov, I. I., Rozanov, A. G., Zhabina, N. N. and Yagodinskaya, T. A. (1972) Sulfur in sediments of the Pacific to the east of Japan, *Litologia i Poleznye Iskopaemye,* **4**, 50-64 (in Russian).

(1974) *World Water Budget and Water Resources of the Earth,* Leningrad, Hydrometeoizdat, (in Russian).

SECTION II

Interactions Between Major Biogeochemical Cycles

Some Perspectives of the Major Biogeochemical Cycles
Edited by Gene E. Likens
© 1981 SCOPE

CHAPTER 5

Chemical Coupling of the Nitrogen, Sulphur, and Carbon Cycles in the Atmosphere

D. H. EHHALT

Institute of Atmospheric Chemistry, Kernforschungsanlage, 5170 Jülich, Federal Republic of Germany

ABSTRACT

The various atmospheric cycles of nitrogen, sulphur, and carbon are coupled through reactions between molecules belonging to the different cycles. It is argued from a simple tropospheric model that much of the coupling proceeds indirectly through the common dependence on the OH-radical which for most trace gases initiates the most efficient removal. In the case of the nitrogen- and carbon-cycles this coupling can be quite strong.

5.1 INTRODUCTION

Traditionally atmospheric trace gas cycles were treated separately. The mass balance was determined from the direct sources and from the direct sinks, disregarding the influence on the other trace gas cycles. The same was done in predictions of increased future concentrations due to man-made pollutants. Recently, however, our understanding of atmospheric chemistry has greatly improved. We have begun to recognize that there are a number of reactions between molecules from different trace gas cycles. As a result it has been realized that there are many more pathways of direct or indirect interaction than previously assumed, that at least some of the cycles cannot be treated independently, and that a much more complicated system analysis is required than hitherto practised. In the following I will try to investigate, if and how much the atmospheric cycles of S, N, and C are mutually coupled.

5.2 THE MAJOR TRACE GASES IN THE ATMOSPHERIC CYCLES OF N, S, AND C

The major source species by which reactive N, S, and C compounds are introduced into the atmosphere are listed in Table 5.1. The major carbon species listed in this

Table 5.1 The Reaction of Atmospheric Trace Gases containing C, N, and S with OH

Trace gas	Mixing ratio in the northern hemisphere (p.p.b.)*	Mean tropospheric lifetime	Contribution of OH-sink reaction (%)
CH_4	1600	5 yr	100
CO	250	60 days	100
Non-methane hydrocarbons (NMHC), C_2-C_5	2-10	1-100 days	50-100
SO_2	0.2	14 days	50
COS	0.5	1 yr	100
H_2S		4 days	100
$(CH_3)_2S$		1 day	50
NO, NO_2	0.1	1 day	100
NH_3	ca. 1	14 days	10
N_2O	310	100 yr	0

*Here b, billion, is 10^9

table are CH_4 and CO. The mixing ratios given refer to unpolluted, ground-level air in the northern hemisphere. CO_2 is not discussed, because, as far as tropospheric gas-phase chemistry is concerned, CO_2 is inert. The non-methane hydrocarbons, C_2 to C_5, are lumped together; the mixing ratio given refers to the total.

Among the sulphur species, SO_2 is considered the most important. Its background mixing ratio is only 0.2 p.p.b. although in polluted areas much higher values have been found. COS is a molecule whose presence in the atmosphere has been demonstrated only recently. It seems to be fairly uniformly distributed with a mixing ratio of about 0.5 p.p.b. which appears consistent with its rather long lifetime of one year. The remaining two S-species, H_2S and $(CH_3)_2S$—and I might have added other mercaptans—are known to be emitted into the atmosphere. However, their lifetimes are so short that their background concentrations are low, highly variable, and not reliably established. Except for the fact that their gas-phase oxidation is started by the reaction with the OH-radical, the homogeneous chemistry of atmospheric sulphur compounds is not well established; what is known has been summarized by Graedel (1977). The most recent scheme for SO_2 oxidation has been suggested by Davis et al. (1979). Rainout has been discussed by Gravenhorst et al. (1978).

The N-cycle also has several subcycles. Because of their multiple interactions, NO and NO_2 are most important chemically. They interconvert rapidly and their combined mixing ratio in background air is about 0.1 p.p.b. They are eventually oxidized to HNO_3 which is present at about 0.1 p.p.b. in the troposphere. NH_3 is present with a mixing ratio of the order of 1 p.p.b. Again, because of the short lifetimes, the mixing ratios of these N-compounds are highly variable and the average background concentration is not well established. Values for N_2O are given in Table 5.1 because of its prominence in the nitrogen group; however like CO_2 it is essentially chemically inert in the troposphere.

There are several features common to the trace gases in Table 5.1.

(i) The major sources are located at the earth's surface, i.e. external to the atmosphere. Therefore, as far as the atmosphere is concerned, they can be treated as independent boundary conditions. In other words, the production rates of these gases are not coupled. The only exception is CO a significant part of which is produced within the atmosphere from the oxidation of CH_4.

(ii) The major part of the cycles takes place within the troposphere (excepting N_2O).

(iii) Much of the chemical conversion into other species is initiated by the OH-radical. The fraction is given in the last column of Table 5.1. Most values lie between 50 and 100 per cent.

(iv) In the case of NO_x and the sulphur species, this conversion leads eventually to acidic and highly soluble species: HNO_3 and H_2SO_4. These reactions contribute greatly to the fact that all of the eventual atmospheric removal in the S-cycle and most of the removal in the N-cycle proceeds through dry and wet deposition.

5.3 THE COUPLING OF THE CYCLES

As indicated in Table 5.1, and as far as we know, reaction with OH provides most of the coupling between the cycles. This coupling is indirect because it proceeds through a perturbation of the OH concentration. Each of the cycles, but particularly the N-cycle and the C-cycle, exerts a certain control over the OH concentration. For example, an increase in the atmospheric concentration of NO_2 leads to an increase in the OH concentration. Since CH_4 or CO are destroyed by OH, their atmospheric concentrations will decrease in inverse proportion. An increase by a factor of two in OH will decrease the atmospheric steady-state concentration of CH_4 by a factor of two.

There are of course other reactions which lead to direct coupling of the cycles. I shall briefly consider two of them. One is provided by reactions of the NH_2 radical with NO and NO_2 as outlined by the reaction scheme of Table 5.2. These data suggest the interesting possibility of a destruction of NH_3 and NO within the troposphere to form N_2 and N_2O as final products. This possibility very much depends on the rate constant of reaction (2), for which so far only an upper limit of 10^{-18} cm^3 molecule^{-1} s^{-1} exists for the effective second order rate constant (Lesclaux and Demissy, 1977). If reaction (2) is much slower than that, reaction (3) and (4) will be the significant removal paths of NH_2 leading to a considerable production of N_2O, up to 6×10^{12} g yr^{-1}, within the troposphere.

Another coupling proceeds through the rainout of soluble gases. Solution of NH_3 in cloud droplets will increase the pH of these droplets and allow more rapid solution of SO_2. Thus the respective wet removal rates are coupled to a small extent.

Table 5.2 Reactions and Reaction Rate Constants of NH_3 and NH_2

Reaction			Rate constant at 298 °K and 1 atm (cm^3 molecule^{-1} s^{-1})
(1) $NH_3 + OH$	→	$NH_2 + H_2O$	1.6×10^{-13}
(2) $NH_2 + O_2 + M$	→	$NH_2O_2 + M$	$< 10^{-18}$
(3) $NH_2 + NO$	→	$N_2 + H_2O$	1.2×10^{-11}
(4) $NH_2 + NO_2$	→	$N_2O + H_2O$	1.2×10^{-11}

(1) Smith and Zellner, 1975; (2) Lesclaux and Demissy, 1977; (3) Hack et al., 1978; (4) Hack et al., 1978.

First estimates indicate, however, that on a global scale these and other conceivable couplings are small compared to that provided by the OH-radical. Therefore the following discussion will focus on the latter.

The coupling of the N, S, and C cycles through OH is by no means fully understood. Nevertheless a sufficient number of reaction paths have been identified to present a rather complicated picture. To simplify matters, the OH-reaction scheme shown in Figure 5.1 has been reduced to the most important reactions and contains only one major representative of each cycle. It applies to background conditions as most of the calculations published so far refer to the background atmosphere. The concentrations, reactions, and reaction rate constants involved are given in Tables 5.3 and 5.4.

As indicated in Figure 5.1, the primary production of radicals proceeds via the photolysis of O_3. At wavelengths below 310 nm O_3 is photolysed to give an excited oxygen atom, O^1D, which partly reacts with H_2O to form OH. OH rapidly reacts with a large number of molecules, it is so to speak the major 'cleansing agent' of the atmosphere. During its reaction it interconverts to other radical species. Through the reaction with O_3, in which O_3 is converted to O_2, OH is converted to HO_2. The attack on CH_4 by OH entails a whole sequence of reactions symbolized by the circle around CH_4 (Figure 5.1), in which CH_4 is decomposed first to H_2CO then to CO and H_2, whereas OH is eventually converted to HO_2. (The reaction of H_2CO with OH is included in the conversion rate given in the arrow; there is slightly more HO_2 produced than OH used, as indicated by the factor α. In the lower troposphere $\alpha \approx 1$.) The fastest conversion of OH, however, is the reaction with CO. The products are CO_2 and atomic H. The H atom reacts so quickly with O_2 to form HO_2 that the H concentration remains very low. The HO_2 formed is less reactive than OH. The back reactions from HO_2 to OH have smaller rate constants and the HO_2 concentration builds up to higher values. The back reactions with O_3 and NO lead directly to OH. HO_2 also reacts with itself to form H_2O_2 which photolyses to form OH. H_2O_2 can also be removed by rainout, heterogeneous reactions or reaction with OH, which leads to a net loss of radicals. Note that interconversion

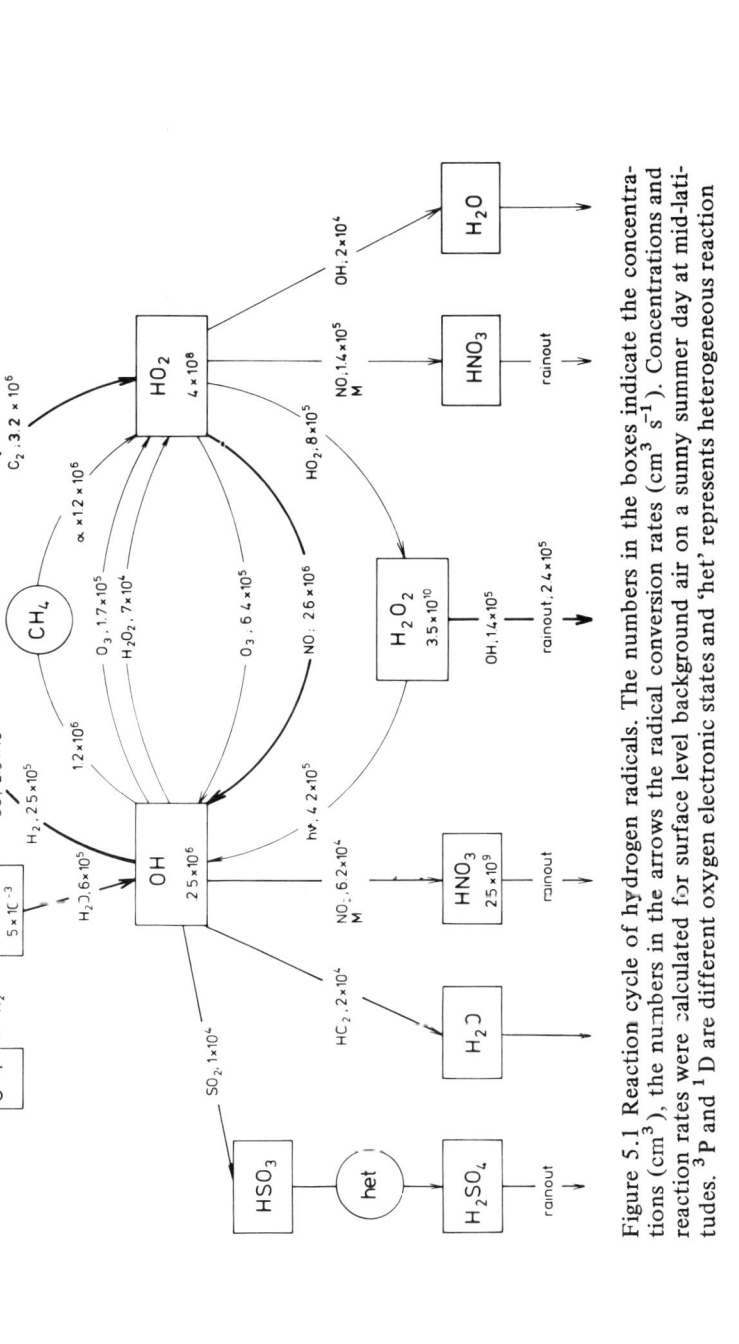

Figure 5.1 Reaction cycle of hydrogen radicals. The numbers in the boxes indicate the concentrations (cm^3), the numbers in the arrows the radical conversion rates (cm^3 s^{-1}). Concentrations and reaction rates were calculated for surface level background air on a sunny summer day at mid-latitudes. ^{3}P and ^{1}D are different oxygen electronic states and 'het' represents heterogeneous reaction

Table 5.3 Average Daytime Concentrations used to derive Figure 5.1

Molecule	Concentration (molecule/cm^3)	Mixing ratio (p.p.b.)
NO	0.8×10^9	0.33×10^{-2}
NO_2	1.7×10^9	0.66×10^{-2}
HNO_3	2.5×10^9	0.1
SO_2	5.0×10^9	0.2
O_3	1.0×10^{12}	40
CO	3.7×10^{12}	150
H_2	13.8×10^{12}	550
CH_4	4×10^{13}	1600
H_2O	2.5×10^{17}	10^7

Table 5.4 Reactions and Reaction Rate Constants* used to derive Figure 5.1

Reaction		Reaction constants (298 °K, 1 atm)
$O^1D + N_2$	$\rightarrow$ $O^3 + N_2$	3×10^{-11} cm^3 molecule^{-1} s^{-1}
$O^1D + H_2O$	$\rightarrow$ $OH + OH$	2.3×10^{-10} cm^3 molecule^{-1} s^{-1}
$OH + H_2$	$\rightarrow$ $H_2O + H$	7.5×10^{-15} cm^3 molecule^{-1} s^{-1}
$OH + CO$	$\rightarrow$ $CO_2 + H$	3×10^{-13} cm^3 molecule^{-1} s^{-1} †
$OH + CH_4$	$\rightarrow$ $CH_3 + H_2O$	7.7×10^{-15} cm^3 molecule^{-1} s^{-1}
$OH + O_3$	$\rightarrow$ $HO_2 + O_2$	6.8×10^{-14} cm^3 molecule^{-1} s^{-1}
$OH + NO_2 + M$	$\rightarrow$ $HNO_3 + M$	1.45×10^{-11} cm^3 molecule^{-1} s^{-1}
$OH + HNO_3$	$\rightarrow$ $H_2O + NO_3$	8.5×10^{-14} cm^3 molecule^{-1} s^{-1}
$OH + H_2O$	$\rightarrow$ $H_2O + O_2$	4.0×10^{-11} cm^3 molecule^{-1} s^{-1}
$OH + SO_2 + M$	$\rightarrow$ $HSO_3 + M$	8×10^{-13} cm^3 molecule^{-1} s^{-1}
$OH + H_2O_2$	$\rightarrow$ $HO_2 + H_2O$	8.1×10^{-13} cm^3 molecule^{-1} s^{-1}
$H + O_2 + M$	$\rightarrow$ $HO_2 + M$	1.4×10^{-12} cm^3 molecule^{-1} s^{-1}
$HO_2 + HO_2$	$\rightarrow$ $H_2O_2 + O_2$	2.5×10^{-12} cm^3 molecule^{-1} s^{-1}
$HO_2 + O_3$	$\rightarrow$ $OH + O_2 + O_2$	1.6×10^{-15} cm^3 molecule^{-1} s^{-1}
$HO_2 + NO$	$\rightarrow$ $OH + NO_2$	8×10^{-12} cm^3 molecule^{-1} s^{-1} ‡
$HO_2 + NO + M$	$\rightarrow$ $HNO_3 + M$	1.4×10^{-13} cm^3 molecule^{-1} s^{-1}
$O_3 + h\nu$	$\rightarrow$ $O_2 + O^1D$	4×10^{-6} s^{-1}
$H_2O_2 + h\nu$	$\rightarrow$ $OH + OH$	6×10^{-6} s^{-1}
H_2O_2	$\rightarrow$ rainout, heterogeneous reactions	3.4×10^{-6} s^{-1}

*Unless otherwise stated the rate constants are taken from the compilation of DeMore et al., 1979; †includes three body reaction; ‡Cox, 1975.

does not mean a net loss of radicals; OH is only temporarily stored as HO_2, which is eventually recycled to OH. Net radical losses are only maintained by the arrows pointing downward and out of the cycle. The loss via H_2O_2 was already mentioned. HO_2 also reacts with OH to form H_2O. In addition, Figure 5.1 shows a not well-established reaction of HO_2 with NO to form HNO_3. This reaction is included because it has been used by Hameed et al. (1979) to calculate the curves shown in Figures 5.2 and 5.3. Another important radical loss is the reaction of OH with NO_2 which yields HNO_3. Finally the reaction of SO_2 with OH is important, because it removes about 50 per cent of the SO_2 from the atmosphere by reaction to HSO_3 followed by heterogeneous reaction to H_2SO_4 and finally by rainout. Owing to the small radical flux, however, it is of very limited influence on the OH balance.

Adding up all the flux values in the arrows, it is seen that the recycling of OH is about 10 times faster than the net loss through the removal reactions, and that OH takes about 0.5 s to pass through one such cycle.

Now I shall identify the coupling points, where the radicals react with the N-, C-, and S-species. Most easily understood is the reaction with SO_2. It leads to a net loss of OH. The loss is small, however, at least in the background atmosphere. There, SO_2 accounts for about one per cent of the net radical loss. Even doubling of the SO_2 concentration would have little impact on the OH concentration and hardly any influence on the C and N cycle.

For the carbon species CO and CH_4 the coupling is more important but also more complicated. Especially, the concentration of CO, whose radical conversion is quite fast, has a large impact on the OH concentration. This impact, however, since it takes place within the interconversion loop, depends also on the rate of the back reaction which is mediated mainly by NO. Thus the impact of CO will also depend on the NO concentration. If the NO concentration is low, the back reaction of HO_2 to OH is slow. An increase in the CO concentration will accelerate the reaction of OH to HO_2 and decrease the OH concentration. If NO concentrations are large, the following feedback comes into play. In the reaction of HO_2 with NO the latter is oxidized to NO_2. NO_2 is then photolysed by solar ultraviolet radiation to give O-atoms which in turn react to form O_3. The O_3 concentration increases, and with it the primary production of OH. This counterbalances the increased conversion rate of OH due to the increased CO concentration. Even more complicated is the situation with respect to NO and NO_2, because they enter into the interconversion, into the net loss of HO_2 and OH, and into the primary production through the feedback via O_3 as just shown. All I can say from the analysis of Figure 5.1 is that an increase of NO_x will increase significantly the net losses, and thus decrease the sum of OH and HO_2. To obtain a more quantitative understanding it is necessary to turn to model calculations of atmospheric chemistry. As an example I shall discuss the recent calculations by Hameed et al. (1979). These calculations were made with a steady-state, zero-dimensional model of the global troposphere. This means that the data I will use to quantify the coupling between the cycles through OH still come from a rather simplified model and do not warrant far-reaching conclusions. The changes of the OH, HO_2, and HO_x concentrations following a change in the global NO_2 concentration are illustrated in Figure 5.2. All

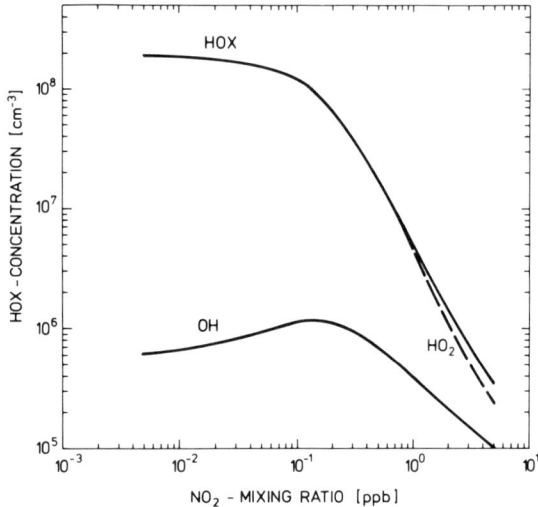

Figure 5.2 The concentration of OH, HO_2, and their sum, HO_x, as a function of the NO_2 mixing ratio calculated from a zero-dimensional, steady-state model of the troposphere (Hameed et al., 1979)

other input parameters remain fixed. For very small initial NO_2 concentrations, 10^{-2} to 10^{-1} p.p.b., the sum of OH and HO_2, HO_x, decreases little with increasing NO_2. In this range the other net losses of radicals, mainly formation and rainout of H_2O_2, apparently still dominate. Only above a NO_2 concentration of about 0.1 p.p.b., do reactions of OH with NO_2, and HO_2 with NO become the dominant net loss terms and the decrease of HO_x with further increase of NO_2 becomes very marked. It can also be seen that over much of the NO_2 range shown in Figure 5.2 the HO_x decrease is essentially due to a decrease of HO_2. In fact, at low NO_2 concentrations OH increases with increased NO_2 concentration. Below about 0.1 p.p.b. NO_2, the OH response to an increased NO_2 concentration is dominated by faster conversion from HO_2 to OH via NO which is in rapid equilibrium with NO_2 and not by the slight increase in the net losses of odd hydrogen radicals from reaction with NO and NO_2. Only above 0.1 p.p.b. NO_2, when the external radical loss due to NO and NO_2 starts to dominate the other external loss reactions, does OH also begin to decrease with increasing NO_2, despite the increased conversion rate of HO_2 to OH.

Obviously the OH response of the system considered depends strongly on the initial NO_2 concentration. In the background troposphere with relatively low NO_2 concentrations (below 0.1 p.p.b.) the response to an increase in NO_2 should be relatively small judging from the flat slope of the OH curve around that concentration. In polluted areas with already high NO_2 concentrations a strong decrease in OH might be expected with further increase in NO_2.

Chemical Coupling of the Nitrogen, Sulphur, and Carbon Cycles 89

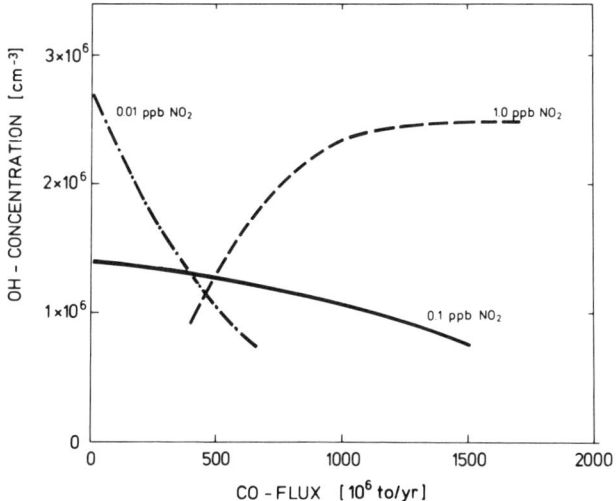

Figure 5.3 The concentration of OH as a function of the CO emission into the troposphere for various mixing ratios of NO_2 (adapted from Hameed et al., 1979)

The change of OH concentration with change in CO emission is shown in Figure 5.3. This change is calculated from the CH_4 response curves given by Hameed et al. (1979) assuming that the only loss for CH_4 is the reaction with OH and that the CH_4 input flux is fixed. Clearly the OH response to an increased CO flux depends on the NO_2 concentration. OH drops rapidly with increasing CO flux for very small NO_2 concentrations, say 0.01 p.p.b., as pointed out in the discussion of Figure 5.1. The response curve for 0.1 p.p.b. NO_2 is much weaker but still decreasing. For large concentrations of NO_2, say 1 p.p.b., an increased emission of CO leads to more OH. This change in the response can be traced to the faster cycling between OH and HO_2: the forward reaction of OH to HO_2 is accelerated by increased CO concentration; the back reaction of HO_2 to OH is accelerated by the increased NO concentration. Since in its reaction with HO_2, NO is oxidized to NO_2 which in turn is rapidly photolysed to NO and O atoms, the faster cycling leads to an increased production of O atoms. The O atoms combine with O_2 to give O_3, and the increased levels of O_3 lead to an increased primary production of OH.

Incidentally, the present anthropogenic emission of CO is about 500×10^6 ton yr^{-1} which gives an OH concentration of about 10^6 OH cm^{-3} for all three curves. Nevertheless a change of OH due to a change in CO could be fairly substantial. It depends on the initial concentrations of CO as well as the NO_2 present in the atmosphere.

These findings are summarized in Table 5.5 and show the relative change in the global concentration due to a change in CO emission or in NO_2 and SO_2 concentration. The numerical value of -1.1 means that the global OH concentration would decrease by 11 per cent if the global CO emission increased by 10 per cent. Obviously,

Table 5.5 Relative Change of OH Concentration with Change in CO, NO_2, and SO_2 Concentrations as a Function of the NO_2 Concentration

Relative change		NO_2 concentration (p.p.b.)			Remarks
		0.01	0.1	1.0	
$\dfrac{\Delta OH}{OH}$	$\dfrac{\Delta CO}{CO}$	-1.1	-0.13	$+1.42$	change of CO flux around 500×10^6 ton yr^{-1} after Hameed et al., 1979
$\dfrac{\Delta OH}{OH}$	$\dfrac{\Delta NO_2}{NO_2}$	$+0.11$	$+0.22$	-0.92	change of NO_2 concentration, after Hameed et al., 1979
$\dfrac{\Delta OH}{OH}$	$\dfrac{\Delta SO_2}{SO_2}$	-0.01	-0.01	-0.01	change of SO_2 concentration around 0.2 p.p.b., estimated from Figure 5.1

depending on the NO_2 concentration, the OH changes due to perturbations of CO or NO_2 can be fairly large. For SO_2 the influence on OH concentrations is small in any case. The middle column for 0.1 p.p.b. NO_2 probably represents the closest estimate of today's global conditions. Fortuitously, the present global response of OH to a perturbation of CO and NO_2, should be relatively small that is, about two per cent for a 10 per cent change in NO_x.

The conclusion from these model calculations then is: Yes, the atmospheric trace gas cycles should be coupled through the OH radical. As man or nature change the carbon or nitrogen cycle, the global OH concentration will change also. The resulting OH change is a complicated function of initial conditions and size of the perturbation. Since reaction with OH is the major sink mechanism for most trace gases, their chemical lifetime in the atmosphere will adjust accordingly. A change in one trace gas cycle, however, will not influence the overall mass balance of the atmospheric cycles of the other trace gases. The mass balance is fixed by the production of the trace gases which except for CO occurs outside the atmosphere and thus to a first approximation remains unaffected. What will change is the atmospheric concentration and global distribution of the other trace gases. It seems, however, judging from Table 5.5, that on a global scale and in the near future the effects due to coupling through OH should be relatively small.

Regional pollution is a different matter. Man introduces and concentrates a large part of the pollutant trace gases in industrial areas. There the perturbation of the local chemical system may become severe and coupling via the OH-radical may be quite important. (Other, direct coupling mechanisms could begin to play a role too.) As Figures 5.2 and 5.3 show, it is conceivable that in such a situation—given the proper mix of pollutants—the local OH concentration could decrease with higher emissions in the future. In that case the self cleansing ability of the local atmosphere would become overloaded. As a result an increasing fraction of the pollutants would spill over into the background atmosphere. In this way regional coupling of the trace gas cycles may produce a global impact. The simple model data presented here can only identify the questions, they do not provide a final answer. In fact much future research needs to be done before the extent of chemical coupling in the atmosphere is understood in reasonable detail.

5.4 REFERENCES

Cox, R. A. (1975) The photolysis of gaseous nitrous acid—a technique for obtaining kinetic data on atmospheric photo-oxidation reactions, *Int. J. Chem. Kinetics,* Symp. 1, 379–398.

Davis, D. D., Ravishankara, A. R. and Fischer, S. (1979) SO_2 oxidation via the hydroxyl radical: Atmospheric fate of HSO_x radicals, *Geophys. Res. Lett.,* **6**, 113–116.

DeMore, W. B., Stief, L. J., Kaufman, F., Golden, D. M., Hampson, R. F., Kurylo, M. J., Margitan, J. J., Molina, M. J. and Watson, R. T. (1979) *Chemical Kinetic and Photochemical Data for use in Stratospheric Modelling,* Eval. 2, NASA Panel for Data Evaluation, JPL Publication 7/9–27.

Graedel, T. E. (1977) The homogeneous chemistry of atmospheric sulfur, *Rev. Geophys. and Space Phys.,* **15**, 421–428.

Gravenhorst, G., Janssen-Schmidt, Th. and Ehhalt, D. H. (1978) The influence of clouds and rain on the vertical distribution of sulfur dioxide in a one-dimensional steady-state model, *Atmospheric Environment,* **12**, 691–698.

Hack, W., Schacke, H., Schröter, M. and Wagner, H. G. (1978) *Reaction Rates of NH_2-radicals with NO, NO_2, C_2H_2, C_2H_4 and Other Hydrocarbons,* Report Max-Planck-Institute für Strömungsforschung, Göttingen.

Hameed, S., Pinto, J. P. and Stewart, R. W. (1979) Sensitivity of the predicted CO-OH-CH_4 perturbation to tropospheric NO_x concentrations, *J. Geophys. Res.,* **84**, 763–768.

Lesclaux, R. and Demissy, M. (1977) On the reaction of NH_2 radical with oxygen, *Nouv. J. de Chimie,* **1**, 443–444.

Smith, I. W. N. and Zellner, R. (1975) Rate measurements of OH by resonance absorption, IV, Reactions of OH with NH_3 and HNO_3, *Int. J. Chem. Kinetics,* Symp., 1, 1–351.

Some Perspectives of the Major Biogeochemical Cycles
Edited by Gene E. Likens
© 1981 SCOPE

CHAPTER 6

Interactions Between Major Biogeochemical Cycles in Terrestrial Ecosystems*

G. E. LIKENS

Section of Ecology and Systematics, Division of Biological Sciences, Cornell University, USA

F. HERBERT BORMANN

School of Forestry and Environmental Studies, Yale University, USA

N. M. JOHNSON

Department of Earth Sciences, Dartmouth College, USA

ABSTRACT

Best approximations at present would suggest that global cycles of carbon, sulphur, nitrogen, and phosphorus have been altered by human activity. However, little quantitative information exists on the interplay between the individual biogeochemical cycles of elements in terrestrial ecosystems. Because of man's potential to modify major biogeochemical cycles, such information is badly needed. Interactive effects may outweigh effects measured in terms of individual elements.

Boundary conditions generally have been ignored in attempts to quantify the flux and cycling of water and chemical elements for terrestrial ecosystems. Watershed (catchment) areas are convenient, functional units of the landscape for such studies.

Relative to watersheds, the hydrologic cycle assumes paramount importance through its effect on all other biogeochemical cycles; e.g. by regulating the timing and amount of inputs of chemicals in precipitation, and loss of dissolved and particulate substances by erosion, in runoff, or deep seepage. In turn, transpiration by plants affects the hydrologic cycle, thereby reducing the loss of nutrients from a drainage area.

*This is a contribution to the Hubbard Brook Ecosystem Study. Financial support was provided by the National Science Foundation. We appreciate a critical review and comments from B. Peterson.

Likewise, acid precipitation, an anthropogenic manifestation of the sulphur and nitrogen cycles, increases the flux of aluminium, phosphorus, calcium, etc., (in or from soils) and may decrease the flux of carbon by inhibiting photosynthesis and decomposition of organic matter. Conversely, as plant nutrients, the nitrogen and sulphur contained in acid precipitation may stimulate primary productivity and decomposition in some ecosystems, and thus enhance the flux of carbon in terrestrial ecosystems. Such interplay between cycles is undoubtedly common, but the quantitative effects are essentially unknown.

Biogeochemical cycles can also be altered indirectly through effects of haze, ozone, smog, CO_2, and dust on climate. In this regard, the carbon cycle is often crucial for interactions between cycles because of the pivotal role of photosynthesis and decomposition.

Comparative studies or experimental manipulations of entire ecosystems are costly and difficult, but the results are very informative. Such studies provide quantitative data on changes in flux and reveal linkages between cycles that are needed to advance the state-of-the-art and to elucidate mechanisms of interaction.

6.1 INTRODUCTION

To date most of the attention relative to biogeochemical cycles has been given to fluxes and mass balances ('budgets') of individual elements such as carbon, nitrogen, sulphur, and phosphorus. Feedback or synergistic type interactions between the cycles of the various elements has received little serious attention (cf. Bolin, 1976; Lerman, et al., 1977). It is technically difficult to monitor and quantify the various fluxes, standing stocks, and processes regulating these parameters for only one element, let alone to study various cycles interacting together within an ecosystem or planetary context. Nevertheless, interactions exist and we shall attempt to point to some of them here. Our examples will be largely drawn from experience in temperate, humid, forested ecosystems.

6.2 TERRESTRIAL ECOSYSTEMS

A critical conceptual consideration in evaluating biogeochemical cycles for terrestrial ecosystems is scale. What are the boundaries of a forest or grassland? Obviously quantitative determinations of fluxes, mass balances, or interactions cannot be made without establishing boundaries, but such pragmatic edges of a terrestrial ecosystem are usually established for the convenience of the investigation rather than to delimit integral functional units.

Lakes, streams, or islands appear to have obvious boundaries—but do they? What are the boundaries? How are they affected during periods of flooding or drought? Aquatic invertebrates may be found many metres below the sediment–water interface or horizontally beyond the banks of the stream and 'beneath' the terrestrial ecosystem (e.g. Hynes, 1974). Where then are the boundaries? Terrestrial landscapes often have political, climatic, topographic, vegetational, or other boundaries. Are these in any way functional boundaries?

The size of 'ecosystems' studied in the past has ranged from plots only a few metres in area to the entire planetary surface. Experimental scale has a profound effect on the theoretical context of a functional ecosystem. Larger systems, for example a global one, tend to be 'closed' with respect to mass. As such they may be in a time-invariant state (equilibrium) or a time-dependent state (disequilibrium). In contrast smaller ecosystems tend to be 'open' with respect to mass and may be in a steady-state mode (time invariant) or a non-steady-state mode (time dependent). Our comprehension of the processes operating within an ecosystem (Johnson, 1971) are thus framed by the state of the boundaries for the system (open versus closed) and the reaction pathways (reversible versus irreversible).

Because of their immense theoretical value, thermodynamic parameters (e.g. equilibrium constants, solubility products, phase diagrams) are occasionally used to describe certain natural processes. Implicit in such attempts, however, is that the field conditions must conform to the exacting conditions required by the thermodynamic considerations. More often than not, however, the necessary conditions for thermodynamic equilibrium are not met in the field, and thermal kinetic theory, or some approximation thereof, is required. Kinetic theory represents a reliable body of knowledge equally as powerful as thermodynamics and perhaps more applicable to natural processes.

Clearly it is much easier to make quantitative measurements of the process rates in smaller ecosystems, but do such areas exhibit all of the functional properties of a larger system? What are the critical functional properties? What are the linkages with other systems that affect these properties? Such questions are rarely addressed by ecologists or geochemists.

The error limits associated with measurements of rate processes within very large ecosystems may be so great as to compromise any real understanding of the biogeochemistry of these systems. The problem of signal-to-noise ratios are rarely addressed when reporting estimates of rates for processes in ecosystems.

In our biogeochemical studies of the Hubbard Brook Experimental Forest in the White Mountains of New Hampshire (cf. Likens *et al.*, 1977; Bormann and Likens 1979), we sought a solution to some of these problems by utilizing rather small, discrete watershed (catchment) ecosystems. We looked for units of the landscape that could be isolated on the basis of flowing water in the system. Because water is such an important vector for chemical flux and cycling in natural forested ecosystems, watershed (phreatic) divides would appear to form both functional and reasonable boundaries, amenable to quantitative analysis (Bormann and Likens, 1967; Likens and Bormann, 1972; Likens *et al.*, 1977; Bormann and Likens, 1979). Our model (Figure 6.1) accounts for the quantitative movement of matter across the boundaries of such a watershed ecosystem by the three principal vectors—meteorologic (movement by atmospheric forces), geologic (movement by alluvial or colluvial forces) and biologic (movement by animals). We consider the release and storage of materials inside the boundaries, for example by weathering, or biomass accumulation and release, as internal features of the ecosystem rather than input/output

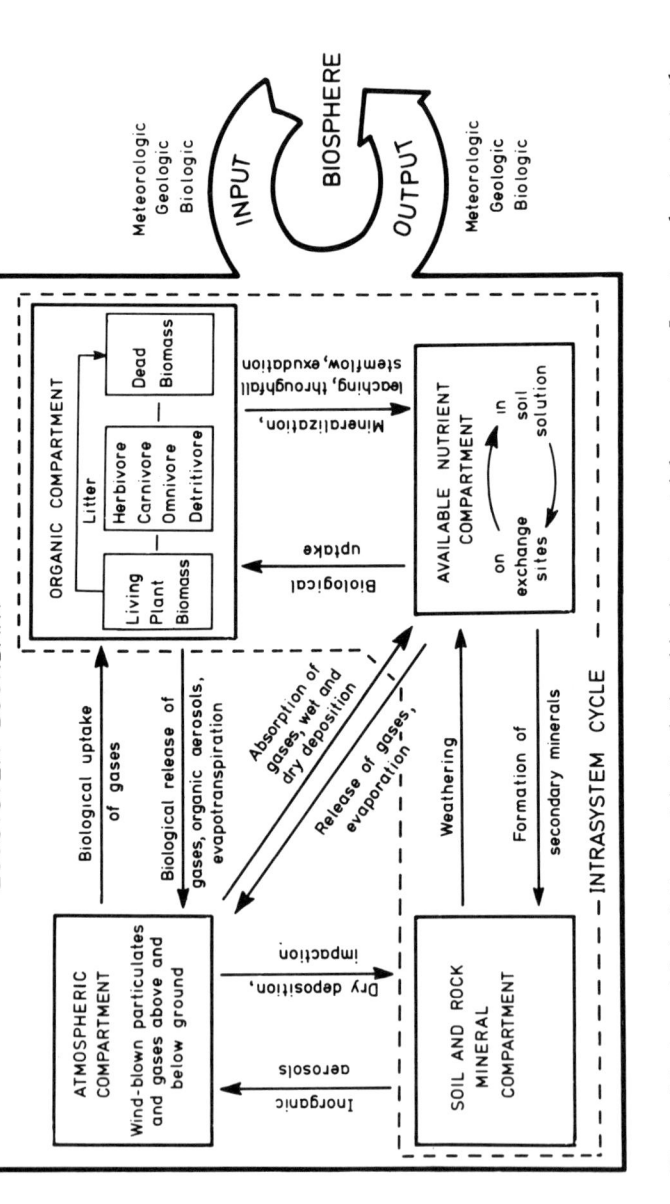

Figure 6.1 A model depicting nutrient relationships in a terrestrial ecosystem. Inputs and outputs to the ecosystem are moved by meteorologic, geologic, and biologic vectors (Bormann and Likens, 1967; Likens and Bormann, 1972). Major sites of accumulation and major exchange pathways within the ecosystem are shown. Nutrients that, because they have no prominent gaseous phase, continually cycle within the boundaries of the ecosystem between the available nutrient, organic matter, and primary and secondary mineral components tend to form an intrasystem cycle. Fluxes across the ecosystem's boundaries link individual ecosystems with the remainder of the biosphere (after Likens et al., 1977)

fluxes but in any event they must be quantitatively accounted for in the bookkeeping of mass balance determinations. All of these vectors and transformations must be carefully considered in quantitative evaluations of biogeochemical cycles. However, surprising deficiencies occur in many evaluations of mass balances for large-scale systems. For example, it is frequently assumed that biological uptake and release of nutrients are in balance (steady state) over a yearly cycle, when this condition probably rarely exists in present-day, man-disturbed landscapes. If biological storage and release do not approach steady state, then errors in mass balance calculations or interpretations can result from simple measurements of inputs in precipitation and outputs in stream water (cf. Likens et al., 1977). To effectively study single cycles or interactions between cycles, it is imperative to use well-defined systems and to make no unwarranted assumptions about the developmental state of the ecosystem. Quantitative studies of watershed ecosystems can provide a conceptual basis for the use and management of landscapes (O'Sullivan, 1979).

6.3 CYCLE INTERACTIONS

It is important to learn at what points the major biogeochemical cycles interact and how these linkages are affected by disturbance. For example, a general question might be—How sensitive is the nitrogen cycle to changes in the sulphur cycle and what are the most sensitive or diagnostic linkages between these two cycles? More specifically it might be asked—How does sulphuric acid in rainfall affect nitrification in the soil, the base exchange capacity, or mineralization activities or microorganisms? Little is known about the answers to such questions, but it is highly likely that as human stress on the natural environment increases, effects on individual biogeochemical cycles will reverberate throughout the system and affect all of the other interlinked cycles. The composite end result may be more important to the overall function of a natural system than an analysis of any single cycle would indicate. It is not clear to us how to obtain such complex answers on a global scale. On the scale of watersheds, however, it is possible by judicious comparative studies or experimental manipulations to elucidate mechanisms, reaction rates, and interactions. Insights may be obtained not only from the results of experimentally imposed disturbance, but it is also possible to 'relax' or remove a pre-existing stress on an ecosystem; e.g. reduce the pollution in a stream or stop the cultivation of a field, and observe the recovery of the system. On a global scale some stress 'experiments' are in process, e.g. increased CO_2 in the atmosphere, acid rain from combustion of fossil fuels, and accelerated use of nitrogenous fertilizers, but unfortunately these alterations were not planned as experiments and the 'controlled replicates' are difficult or impossible to construct. Under these circumstances it is difficult to quantify the effects on individual cycles, let alone complex interactions between cycles. As a result, scientists frequently have resorted to smaller, more manageable systems and/or simulation models with the hope of extrapolating these results to a global scale. At this point, in the development of such information,

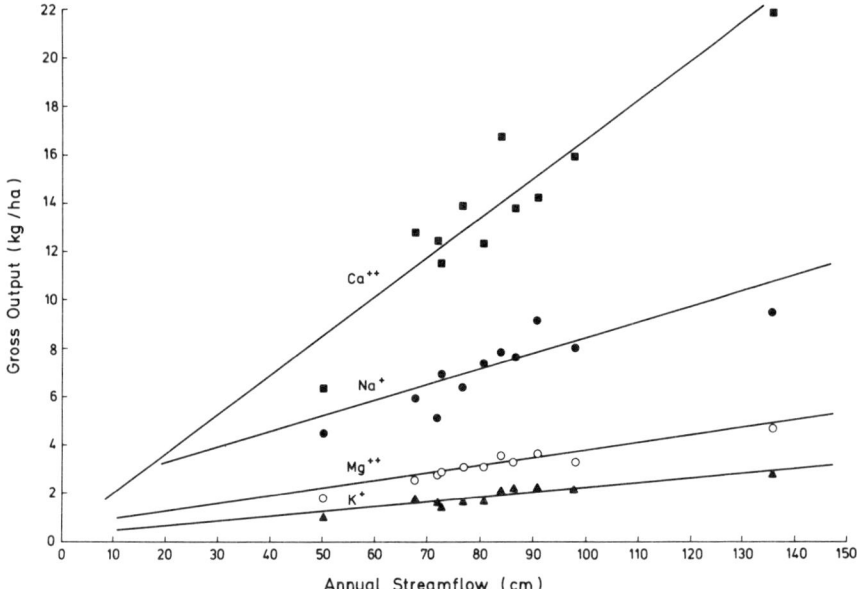

Figure 6.2 Relationship between streamflow and gross output of cations during 1963 to 1974 at the Hubbard Brook Experimental Forest (after Likens et al., 1977)

however, global valuations of individual biogeochemical cycles are at the very best only rough approximations.

6.3.1 Hydrologic Cycle

Water is perhaps the most important linkage mechanism for chemical elements in terrestrial ecosystems. The hydrologic cycle affects other biogeochemical cycles in numerous (direct and indirect) ways. Water acts as a solvent, carrier, catalyst and reagent in many ecosystem reactions and processes. The flux of nutrients into and through many terrestrial ecosystems is largely dependent on the amount and temporal distribution of rain and snow and the chemical content of this water. For example, in an extreme case, Art et al., (1974) found for a coastal terrestrial ecosystem on Long Island, New York, that all of the inputs of calcium, magnesium, sodium, and potassium came from bulk precipitation and deposition of sea spray. Smaller but significant amounts of these nutrients are added to natural ecosystems in bulk precipitation at greater distances from the ocean (cf. Likens et al., 1977). Likewise, losses of major nutrients are usually a function of the amount of water draining from an area (Figure 6.2). In this regard, water lost by transpiration, which is an indirect function of the carbon cycle via photosynthesis, is not available for drainage and thereby transpiration serves to reduce the dissolved nutrient flux out

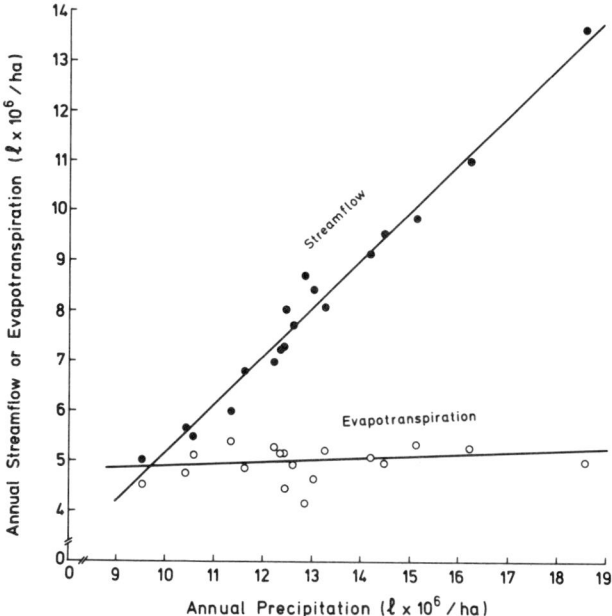

Figure 6.3 Relationship among precipitation, streamflow, and evapotranspiration for the Hubbard Brook Experimental Forest during 1965 to 1974. Transpiration accounts for the majority of the annual evapotranspiration water loss for this humid forest ecosystem (after Likens et al., 1977)

of the system. For example, at the Hubbard Brook Experimental Forest, on average, some 38 per cent of the annual precipitation is lost as evapotranspiration each year (Figure 6.3).

The amount of available water affects the rate of primary production (carbon cycle) in terrestrial ecosystems (e.g. Leith, 1973) and, hence, the rate of uptake and storage of nutrients in biomass. It also plays a major role in the regulation of the processes of throughfall, leaching, decomposition, and mineralization which are integral parts of all nutrient cycles in humid regions.

Water is the primary agent of erosion and transport in humid regions (cf. Bormann et al., 1974; Dunne and Leopold, 1978; Leopold et al., 1964). The downstream transport of particulate matter constitutes an important loss in terrestrial ecosystems, particularly for relatively insoluble elements such as phosphorus, iron, and lead. A major biogeochemical aspect of erosion is related to the removal of exchange surfaces, since not only does this result in the loss of available nutrients, but loss of exchange surfaces can greatly alter the capacity of the system to regulate nutrient cycles and to recover from disturbance. The regulation of the hydrologic cycle and, therefore, erosion is dependent on the developmental phase of the terrestrial ecosystem (cf. Bormann and Likens, 1979), and may be drastically changed by disturbance, man-made or otherwise.

It is noteworthy that where knowledge of hydrologic fluxes is uncertain, associated chemical element budgets are also poorly known. Semi-arid and arid ecosystems exemplify this condition. It is difficult at best to deal quantitatively with the biogeochemistry of ecosystems where the hydrologic budget is unknown, unreliable or ephemeral.

In summary, it is difficult to exaggerate the importance of the water cycle with regard to practically every aspect of biogeochemistry. Massive irrigation works, diversions of major river systems, large-scale cloud seeding, major impoundments and deforestation or urbanization of catchment areas are all examples of man's conscious or unconscious efforts in manipulating the hydrologic cycle. All of the above have well-documented effects in dislocating or disturbing the biogeochemical activity within the affected area. Perhaps the ultimate hydrologic trauma, however, occurs when the hydrologic cycle switches from a 'normal' mode to an 'ice age' mode. Can man's influence on the CO_2 balance, atmospheric dust and photochemical smog, and albedo of the earth bring about this most profound and dramatic interaction of all?

6.3.2 Carbon Cycle

The carbon cycle often regulates interactions between various other biogeochemical cycles because of its pivotal role through photosynthesis and decomposition. For example, if ecosystem production exceeds decomposition then storage of nutrients would occur as biomass accumulates. If decomposition predominated, then nutrients would be released from the decaying organic matter. It must be realized, however, that these are generalized responses, since actual uptake and mineralization rates will be dependent on behaviour of different vegetational and chemical species, or both (e.g. Gosz et al., 1973; Gosz, 1981).

Nutrient release through decomposition of leaf tissue in the Hubbard Brook Experimental Forest is influenced by critical carbon to element or element to phosphorus ratios (Gosz et al., 1973). Once the critical C:N, C:P, etc., ratio is attained, mineralization occurs. Threshold ratios may exist for other elements and thereby control their mineralization rates in decomposing organic matter. Thus the presence of one nutrient can affect the availability of another through this type of biogeochemical interaction.

Currently a great amount of attention is being focused on the carbon cycle because of increasing concentrations of CO_2 in the atmosphere from the combustion of fossil fuels and the resulting potential for climatic change (cf. Anonymous, 1977). Spatially, terrestrial ecosystems are quite heterogenous, i.e. some areas store carbon while others lose it. Globally, however, it is thought that terrestrial ecosystems are a net source of carbon for the atmosphere (see Olson et al., 1978; Bolin,

Table 6.1 Elemental Ratios (by weight) for Major Nutrients

	C	:	N	:	S	:	P
Global[1]							
Fossil fuel emissions	9300	:	36	:	130	:	1
Land plants	790	:	7.6	:	3.1	:	1
Ocean plants	129	:	12	:	2.9	:	1
Soil humus	54	:	3	:	1.2	:	1
Hubbard Brook Experimental Forest							
Leaf material[2]	268	:	13	:	1	:	1
Total living biomass[3]	821	:	6	:	0.6	:	1
Forest floor	358	:	16.5	:	1.6	:	1
Precipitation	775	:	162	:	317	:	1
Runoff	615	:	200	:	800	:	1

[1] Modified from Delwiche and Likens (1977); amount of phosphorus in land plants decreased 4-fold and ratios adjusted accordingly;
[2] weighted for species composition;
[3] weighted for above and below ground materials of different species.

1977; Woodwell et al., 1978). Although the amounts and assumptions underlying this flux are disputed (e.g. Broecker et al., 1979). Moreover, the prevailing wisdom, although not certain either, is that elevated levels of CO_2 in the atmosphere have not, in turn, stimulated terrestrial photosynthesis on a global scale (cf. Woodwell et al., 1978). If net global ecosystem production were enhanced, then storage of nitrogen, phosphorus, and other nutrients in the accumulating biomass would be increased accordingly and transpiration also might be altered.

The so-called 'Redfield ratio' for content of carbon, nitrogen and phosphorus in organic material is widely quoted and used in 'megacalculations' for ecosystem and global budgets. This atomic ratio for C:N:P or 106:16:1 (weight ratio of 40:7:1) was based on analysis of marine plankton (Redfield, 1958; Redfield et al., 1963). Ratios for terrestrial ecosystems, dead organic matter, or global systems may differ widely from the values for marine plankton (Table 6.1), and it is not appropriate to use the Redfield ratio generally for terrestrial ecosystems. The fact that the ratios are different is not at all surprising since the terrestrial plants have a much greater proportion of carbon (cellulose) stored in structural tissues (e.g. wood). Even 'ocean plants' (Table 6.1) have a different ratio than that proposed by Redfield because macrophytes and attached algae are included.

Redfield (1958) argued convincingly that organisms controlled the proportion of inorganic nitrogen and phosphorus in the sea. Furthermore, he proposed strong interaction between these two elements which resulted in a net equilibrium and fixed C:N:P ratio in marine organisms. The same argument can be made for terrestrial systems even though the C:N:P ratios are significantly different from those found in the sea. According to Redfield's arguments, phosphorus is the '. . . master element which controls the availability of the others,' because it is present in lowest supply at the earth's surface relative to carbon, nitrogen, and oxygen. This relatively simple relationship, however, does not necessarily hold for terrestrial ecosystems where nitrogen, water, and other factors frequently limit growth (e.g. Waring and Franklin, 1979).

Simulation models of forest growth may give insight into the effects of vegetation disturbance or effects of accelerated plant growth on biogeochemical cycles (e.g. Botkin *et al.*, 1973; Shugart and West, 1977). For example, simulation suggests that catastrophic losses of major tree species (e.g. blights of chestnut, hemlock from the hardwood forests of the eastern USA) would affect biogeochemical cycles in a variety of ways, e.g. via uptake, release and differential storage of nutrients, gaseous exchange, impaction of aerosols, etc. The carbon cycle would be temporarily affected by such changes, as would be the flux and cycling of N, P, S, and other nutrients.

Harvesting of forest products by clearcutting provides an example of how various cycles may be affected through their linkages. Immediately following clearcutting of northern hardwoods in the northeastern USA, photosynthesis would be reduced and decomposition would be increased. As a result, transpiration would be increased; inventories, accumulation and release of all elements in biomass would be altered depending on the type of harvest and developmental stage of the ecosystem. Nitrification would be accelerated and there would be increased loss of calcium, magnesium, potassium, sodium, nitrate, phosphate, and hydrogen ion and decreased loss of sulphate in stream water (cf. Bormann and Likens, 1979).

Another aspect that should be considered here is the developmental stage of the forest ecosystem (cf. Bormann and Likens, 1979). Carbon flux might be greater for young forests in early stages of development, but impaction of sulphurous, nitrogenous, and phosphorus aerosols probably would be less because the architecture (impaction surface) of the forest would not be fully developed. The effects of ecosystem development and dramatic or subtle shifts in species composition on biogeochemical cycles and their interaction are areas worthy of detailed study (cf. Bormann and Likens, 1979; Gorham *et al.*, 1979).

6.3.3 Sulphur Cycle

Apparently, human activity, mainly through the combustion of fossil fuels, has altered the global sulphur cycle relatively more than for any of the other major elements (Table 6.2). One current estimate suggests that some 65 Tg of gaseous S are emitted to the atmosphere each year by anthropogenic activities, whereas only about 40 Tg-S yr^{-1} are contributed by natural sources (see also Chapters 3 and 4). A manifestation of this alteration is the widespread phenomenon of acid precipitation in Europe and North America (e.g. Likens, 1976; Likens *et al.*, 1972, 1979). Acid precipitation is defined as rain and snow that has a pH of less than 5.6. The cause is strong acids (sulphuric and nitric) originating as combustion products (SO_x,

Table 6.2. Annual Gaseous Emissions to the Atmosphere (based on Delwiche and Likens, 1977; Granat et al., 1976; Svensson and Söderlund, 1976)

	Carbon (Tg)	Sulphur (Tg)	Nitrogen (Tg)
Natural	$\sim 75 \times 10^3$	~ 40	~ 210
Anthropogenic (combustion of fossil fuels)	5×10^3	65	18

NO_x) from fossil fuels. Annual pH values of precipitation average between 4.0 and 4.5 over large areas of western Europe and eastern North America and pH values in the twos and threes are not uncommon for individual storms (Figure 6.4).

Figure 6.4 Distribution of acid precipitation in North America and Europe. Areas designated 10, 20, and 30 X receive 10, 20, and 30 times more acid in precipitation than expected if the pH were 5.6 (after Likens and Butler, 1981)

As rainwater falls through the canopy of a forest, its chemistry is markedly altered (cf. Likens et al., 1977). These changes suggest both an effect on the various biogeochemical cycles by the vegetation and an effect on the vegetation itself (carbon cycle) by the rainfall. Highly acidic precipitation may reduce the primary productivity of terrestrial ecosystems through a variety of mechanisms (direct toxicity, leaching of nutrients from foliage, stress, etc.). As a result the carbon cycle would be affected, and nutrient uptake from the soil as well as accumulation of elements in biomass would be reduced. Conversely, there is evidence that the rate of decomposition also might be slowed as a result of acid precipitation, perhaps having a secondary effect on productivity through a reduction in nutrient supply. Sulphur (and particularly nitrogen and phosphorus) compounds in precipitation may increase productivity if these nutrients are limiting in natural systems, and such increased productivity would alter the uptake, storage, and release of other mineral nutrients. Leaching of nutrients from the leaves could affect predator/prey relations, thereby altering the capacity of the vegetation to resist herbivores or pathogens. In turn, production, nutrient uptake, storage and release, etc., could change.

The potential interaction of the sulphur cycle, via acid precipitation, with other biogeochemical cycles is great. For example, if soil pH were lowered, absorption of ammonia would be enhanced and vice versa (e.g. Allison, 1973). Acid precipitation enhances the solubility and leaching of aluminium and other major cations (Ca, Mg, Na, K) from the soil (cf. Likens et al., 1977; Johnson, 1979). A variety of other elements are also involved with these reactions including dissolved organic carbon, phosphorus, silicon, and fluorine compounds (Driscoll, 1980) thereby affecting their cycles through dissolution, leaching, and redeposition. The linkage between the sulphur–aluminium–phosphorus cycles requires careful study relative to interactions between biogeochemical cycles and their alterations by humans.

A revealing case study is the effect on chemical weathering when acid rain is imposed on the landscape. As seen in the Hubbard Brook Experimental Forest (Johnson, 1979) and in the Adirondack Mountains of New York State (Driscoll, 1980), the intrusion of acid rain into the weathering process has disrupted what is believed to be the normal chemical processes. The strong mineral acids of acid rain appear to have displaced weak acids such as H_2CO_3, which ordinarily cause weathering of mineral substrates. No indications of carbonic acid reactions (carbonation) are evident in the headwater drainage streams from either area. Furthermore, the acid rain dissolves and mobilizes Al^{3+} from the soil zone, where usually it is deposited. The acidified, aluminium-rich drainage waters from these chemical reactions are toxic to fish in downstream waters (Driscoll et al., 1980). Ironically, the acid rain may not itself kill fish, but one of its neutralization products (Al^{3+}) does. This illustrates how a perturbation in one cycle, in this case the sulphur cycle, in one part of an ecosystem, can propagate through the system with unexpected and sometimes undesirable results.

Another dimension of this problem concerns the effect of acid rain on chemical weathering rate. Studies by Granat et al. (1976) suggest that acid rain is accelerating chemical weathering rates substantially, perhaps even doubling the normal rate globally. The assumption is that the effects of acid rain are superimposed upon preexisting reactions, so that the effect is additive. On the other hand, Johnson (1979) believes that, 'no excessive chemical weathering activity can be attributed to acid rain over the northeastern United States,' because of a displacement of the weak acid activity by the strong acids. These contradictory views illustrate the difficulty of comparing global estimates and assumptions with data from discrete ecosystems.

Sulphurous aerosols from human activity can result in increased haziness (e.g. Husar et al., 1979) which, in turn, can affect climate and plant productivity (and all the associated biogeochemical cycles). There is evidence that atmospheric haziness is increasing in the arctic as well as in industrialized areas of the world. Thus, the sulphur cycle could have an affect on the albedo, heat budget, and ultimately the climate of the earth.

The transport and deposition of metals such as zinc, copper, lead, cadmium, etc., from industrial pollution sources has been shown to depress vegetation growth and decomposition rates in forest litter (e.g. Rühling, and Tyler, 1973; Jordan, 1975). Concentrations of some of these metals in rain and snow is surprisingly high and significant accumulation is occurring in remote forested ecosystems of the northeastern USA (Galloway and Likens, 1979; Siccama and Smith, 1978). Thus, it is likely that important feedbacks are occurring between these metal cycles, acid precipitation, and other cycles such as N, S, and P through effects on productivity and decomposition.

Pollution standards or regulations probably show the greatest weakness in using the single factor or single-cycle approach. For example, atmospheric sulphur standards might be based on the role of sulphur in acid precipitation. However, acid precipitation also is the result of atmospheric nitrogen oxides, chlorides, ammonia, metallic bases, etc. Hence control of the acid precipitation problem requires simultaneous consideration of all of these elements and their interactions.

6.3.4 Nitrogen Cycle

Annually some 40 Tg of N are fixed commercially as agricultural fertilizer (Delwiche and Likens, 1977). There has been appreciable concern about the role of this fertilizer in the eutrophication of aquatic ecosystems (e.g. Kohl et al., 1971), and whether gaseous losses from the soil would be enhanced and thereby produce atmospheric imbalances (e.g. McElroy et al., 1977; Crutzen and Ehhalt, 1977; Council for Agricultural Science and Technology, 1976; Hutchinson and Mosier, 1979). E. Lemon and co-workers (personal communication) at Cornell University have found that the vegetational uptake of gaseous NH_3 may serve as a balancing mechanism for levels of ammonia in the atmosphere. Apparently agricultural vegetation can release NH_3 when ambient atmospheric concentrations are low, and absorb it when gaseous concentrations are high. In fact, the uptake rate is accelerated at higher concentrations. Lemon has suggested that if nitrogen were limiting to photosynthesis, then losses of NH_3-N (from fertilizer as well as naturally-fixed N) in one area could stimulate net production by plants in another area via the mechanism described above and thus influence the storage of atmospheric CO_2 in living and dead biomass. In turn, all other major nutrients would be affected proportionally by the changing levels of photosynthetic and decomposition activity.

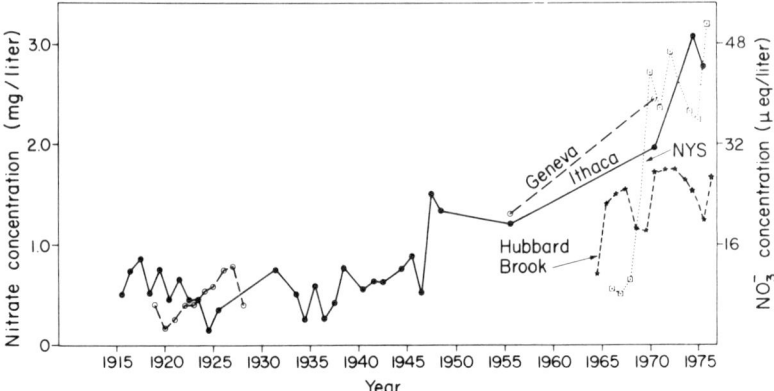

Figure 6.5 Annual nitrate concentrations in precipitation in the northeastern USA (after Likens and Bormann, 1980)

Biologically significant amounts of nitrogen are added to terrestrial ecosystems each year in rain and snow. At various locations in the eastern USA annual inputs in precipitation range between 6 and 10 kg-N ha^{-1} (Likens et al., 1977; Likens, 1972). Wet deposition values in industrialized Europe may be even greater (Söderlund, 1977). Such inputs have increased significantly during the last decade in both the eastern USA (Figure 6.5) and western Europe (e.g. Likens, 1976). There is no quantitative evidence as yet whether these increased inputs have had a significant effect on plant productivity or decomposition in natural ecosystems.

An interesting, but as yet unexplained, finding from the Hubbard Brook Ecosystem Study is the relationship between NO_3^- and SO_4^{2-} in drainage water (cf. Likens et al., 1970). In a recently deforested watershed, it was found that the concentration of NO_3^- in stream water was inversely proportional to the concentration of SO_4^{2-}, to a minimum value of SO_4^{2-} ($\sim$ 3 mg litre^{-1}) which approximated the concentration in ambient precipitation (Figure 6.6). A similar but less pronounced relationship was observed in stream water draining from forested watersheds (Figure 6.7).

6.3.5 Phosphorus Cycle

As discussed above, phosphorus is frequently limiting to plant growth in natural ecosystems. As such the interactions with other biogeochemical cycles, particularly carbon, are obvious.

In most natural ecosystems where budgetary information is available, the input of phosphorus via bulk precipitation exceeds the losses in drainage waters (Likens et al., 1977). This finding would imply that the biomass in these diverse ecosystems is not in steady state and that organic matter was accumulating, or that phosphorus was stored geochemically in the soil. At Hubbard Brook the ratio of annual storage

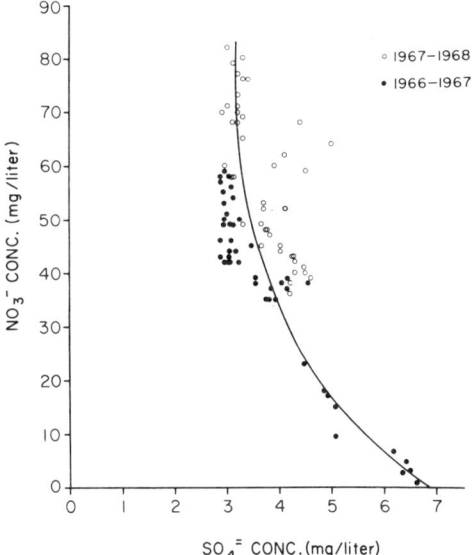

Figure 6.6 Relationship between nitrate and sulphate concentrations in stream water from a deforested watershed of the Hubbard Brook Experimental Forest. Nitrate values less than 25 mg litre^{-1} show the transition period between forested and deforested conditions (after Likens et al., 1970)

in the biomass to input was higher for phosphorus than for any other nutrient in this forested ecosystem (Table 6.3). Any stimulation of net ecosystem production by additions of phosphorus would correspondingly affect many, if not all, of the other elemental cycles.

The origin of phosphorus in rain and snow is largely unknown. Some phosphorus may originate from terrestrial dust swept into the atmosphere, but there are indications that concentrations are higher in rain and snow adjacent to urban and industrial centres.

An excess of phosphorus in aquatic ecosystems usually results in eutrophication (cf. Schindler, chapter 7). However, it is an open question as to how much of this phosphorus originates from fertilizer, applied to adjacent terrestrial ecosystems. In general, forested ecosystems lose smaller amounts of phosphorus in drainage waters than cultivated or urbanized areas, but with disturbance, levels in stream water generally increase (cf. Likens and Bormann, 1974). Also more phosphorus is typically lost in drainage from areas with sedimentary rocks and fertile soil other factors being similar (Dillon and Kirchner, 1975). Much of the phosphorus in surface waters adjacent to urbanized areas comes from the industrial and domestic use of detergents containing phosphorus, and from sewage.

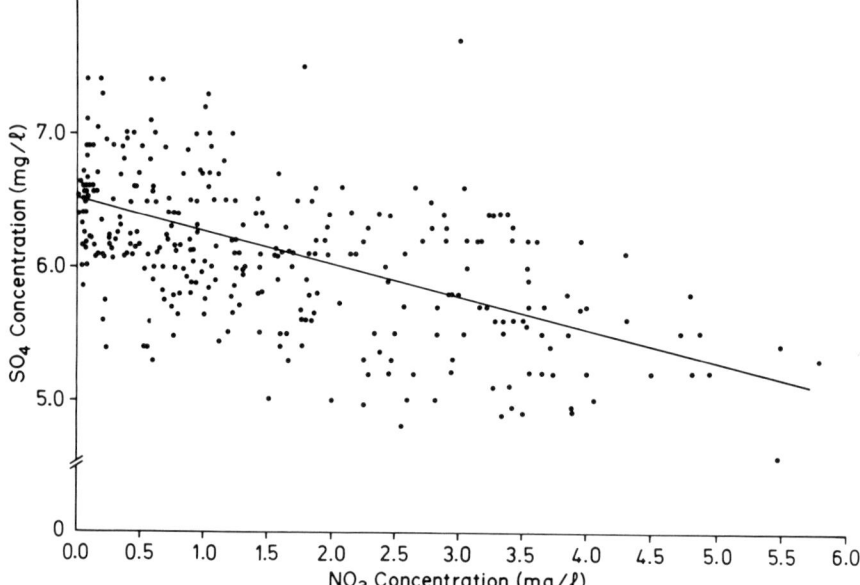

Figure 6.7 Relationship between nitrate and sulphate concentrations in stream water from forested watersheds of the Hubbard Brook Experimental Forest during October to May of 1964 to 1974 (after Likens et al., 1977)

Table 6.3 Annual Input/Output and Fate of Nutrients in the Hubbard Brook Experimental Forest. Values in %

Element	Source		Annual storage/input	Annual output/storage
	Meteorologic input	Weathering release		
Na	22	78	2	4400
S	96	4	10	880
Mg	15	85	22	370
Ca	9	91	41	146
K	11	89	76	39
N	100	<1	81	24
P	1*	99*	99*	0.7

*Estimated

Appreciable amounts of carbon and nitrogen are lost to the atmosphere when natural ecosystems are burned, but it is not known how much phosphorus is lost in this way.

6.4 SUMMARY

Terrestrial cycles of water and all nutrients, including carbon, nitrogen, sulphur, and phosphorus, have been altered by man through such activities as combustion of fossil fuels, application of agricultural fertilizers and pesticides, deforestation, erosion, and irrigation. However, quantitative data are so deficient that it is difficult to determine how much the cycles of these elements have been affected, particularly on a global scale. There is almost no useful quantitative information on interaction between global cycles, but we suspect that interactive effects may outweigh individual effects that we know for single elements.

Not only is there interaction between cycles, but also between systems. It is unrealistic to consider terrestrial systems or cycles in isolation from adjacent aquatic and atmospheric systems and cycles.

Comparative studies or experimental manipulations of entire ecosystems are costly and difficult, but the results can provide quantitative data on processes and reaction rates for biogeochemical cycles, and for interactions between cycles. Quantitative data on flux as well as information on linkages between cycles are needed in a variety of ecosystem types to elucidate mechanisms of interaction and to advance the state of the art. We believe there is a need to establish international technical working groups, primarily concerned with the measurement and prediction of interactions between global cycles. This effort should be an important part of any overall evaluation of the effects of human activities of the global environment.

6.5 REFERENCES

Allison, F. E. (1973) *Soil Organic Matter and its Role in Crop Production*, Amsterdam, Elsevier.

Anonymous (1977) *Energy and Climate*, Washington D.C., National Academy of Sciences.

Art, H. W., Bormann, F. H., Voigt, G. K. and Woodwell, G. M. (1974) Barrier Island Forest Ecosystem: role of meteorologic nutrient inputs, *Science*, **184**, 60–62.

Bolin, B. (1976) Transfer process and time scales in biogeochemical cycles, in Svensson, B. H. and Söderlund, R. (eds), Nitrogen, phosphorus and sulphur – Global cycles, SCOPE Report 7, *Ecol. Bull. (Stockholm)*, **22**, 17–22.

Bolin, B. (1977) Changes of land biota and their importance for the carbon cycle, *Science*, **196**, 613–615.

Bormann, F. H. and Likens, G. E. (1967) Nutrient cycling, *Science*, **155**, 424–429.

Bormann, F. H. and Likens, G. E. (1979) *Pattern and Process in a Forested Ecosystem*, New York, Springer-Verlag.

Bormann, F. H., Likens, G. E., Siccama, T. G., Pierce, R. S. and Eaton, J. S. (1974) The export of nutrients and recovery of stable conditions following deforestation at Hubbard Brook, *Ecol. Monogr.*, **44**(3), 255–277.

Botkin, D. B., Janak, J. F. and Wallis, J. R. (1973) Estimating the effects of carbon fertilization on forest composition by ecosystem simulation, in Woodwell, G. M. and Pecan, E. V. (eds), *Carbon and the Biosphere*, Proc. 24th Brookhaven Symposium in Biology, Upton, New York, 328–344.

Broecker, W. S., Takahashi, T., Simpson, H. J. and Peng, T. H. (1979) Fate of fossil fuel carbon dioxide and the global carbon budget, *Science*, **206**, 409–418.

Council for Agricultural Science and Technology (1976) *Effect of Increased Nitrogen Fixation on Stratospheric Ozone*, Report No. 53, Iowa State University, Ames.

Crutzen, P. J. and Ehhalt, D. H. (1977) Effects of nitrogen fertilizers and combustion on the stratospheric ozone layer, *Ambio*, **6**(2-3), 112–117.

Delwiche, C. C. and Likens, G. E. (1977) Biological response to fossil fuel combustion products, in Stumm, W. (ed), *Global Chemical Cycles and Their Alterations by Man*, Berlin, Dahlem Konferenzen, 33–88.

Dillon, P. J. and Kirchner, W. B. (1975) The effects of geology and land use on the export of phosphorus from watersheds, *Water Res.*, **9**, 135–148.

Driscoll, C. T. (1980) Chemical characterization of some dilute acidified lakes and streams in the Adirondack region of New York State, *Ph.D. Thesis*, Cornell University.

Driscoll, C. T., Baker, J. P., Bisogni, J. J., Jr. and Schofield, C. L. (1980) Aluminium speciation in dilute acidified surface waters of the Adirondack region of New York State, *Nature*, **284**, 161–164.

Dunne, T. and Leopold, L. B. (1978) *Water in Environmental Planning*, San Francisco, W. H. Freeman.

Galloway, J. N. and Likens, G. E. (1979) Atmospheric enhancement of metal deposition in Adirondack lake sediments, *Limnol. Oceanogr.*, **24**(3), 427–433.

Gorham, E., Vitousek, P. M. and Reiners, W. A. (1979) The regulation of chemical budgets over the course of terrestrial ecosystem succession, *Ann. Rev. Ecol. Syst.*, **10**, 53–84.

Gosz, J. R. (1981) Nitrogen cycling in coniferous ecosystems, in *Terrestrial Nitrogen Cycles – Processes, Ecosystem Strategies, and Management Impacts*, SCOPE Report 18, Workshop at Gysinge Wärdshus, Osterfarnebo, Sweden (in press).

Gosz, J. R., Likens, G. E. and Bormann, F. H. (1973) Nutrient release from decomposing leaf and branch litter in the Hubbard Brook Forest, New Hampshire, *Ecol. Monogr.*, **43**(2), 173–191.

Granat, L., Rodhe, H. and Hallberg, R. O. (1976) The global sulfur cycle in Svensson, B. H. and Söderlund, R. (eds), Nitrogen, phosphorus and sulphur – Global cycles, SCOPE Report 7, *Ecol. Bull. (Stockholm)*, **22**, 89–134.

Husar, R. B., Patterson, D. E., Holloway, J. M., Wilson, W. E. and Ellestad, T. G. (1979) Trends of eastern US haziness since 1948, in *4th Symposium on Atmospheric Turbulent Diffusion and Air Pollution*, January 1979, Reno, Nevada.

Hutchinson, G. L. and Mosier, A. R. (1979) Nitrous oxide emissions from an irrigated cornfield, *Science*, **205**, 1125–1127.

Hynes, H. B. N. (1974) Further studies on the distribution of stream animals within the substratum, *Limnol. Oceanogr.*, **19**(1), 92–99.

Johnson, N. M. (1971) Mineral equilibria in ecosystem geochemistry, *Ecology*, **52**(3), 529–531.

Johnson, N. M. (1979) Acid rain: neutralization within the Hubbard Brook ecosystem and regional implications, *Science*, **204**, 497–499.

Jordan, M. J. (1975) Effects of zinc smelter emissions and fire on a chestnut-oak woodland, *Ecology*, **56**, 78–91.

Kohl, D. H., Shearer, G. B. and Commoner, B. (1971) Fertilizer nitrogen: contribution to nitrate in surface water in a corn belt watershed, *Science*, **174**, 1331–1334.

Leopold, L. B., Wolman, M. G. and Miller, J. P. (1964) *Fluvial Processes in Geomorphology*, San Francisco, W. H. Freeman.

Lerman, A. (Rapporteur), Bernhard, M., Bolin, B., Delwiche, C. C., Ehhart, D. H., Gessel, S. P., Kester, D. R., Krumbein, W. E., Likens, G. E., Mackenzie, F. T., Reiners, W. A., Stumm, W., Woodwell, G. M. and Zinke, P. J. (1977) Fossil fuel burning: its effects on the biosphere and biogeochemical cycles, Group Report in Stumm, W. (ed.), *Global Chemical Cycles and Their Alterations by Man*, Berlin, Dahlem Konferenzen, 275-289.

Lieth, H. (1973) Primary production: terrestrial ecosystems, *Human Ecology*, 1(4), 303-332.

Likens, G. E. (1972) The chemistry of precipitation in the central Finger Lakes Region, *Water Resources Center Technical Report 50*, Cornell University, Ithaca, New York.

Likens, G. E. (1976) Acid precipitation, *Chemical and Engineering News*, 54, 29-44.

Likens, G. E. and Bormann, F. H. (1972) Nutrient cycling in ecosystems, in Wiens, J. (ed.), *Ecosystem Structure and Function*, Corvallis, Oregon State University Press, 25-67.

Likens, G. E. and Bormann, F. H. (1974) Linkages between terrestrial and aquatic ecosystems, *BioScience*, 24(8), 447-456.

Likens, G. E. and Bormann, F. H. (1980) The role of watershed and airshed in lake metabolism, in Rodhe, W., Likens, G. E. and Serruya, C. (eds), Proc. Symposium on Lake Metabolism and Lake Management, *Arch. Hydrobiol. Beih. Ergebn. Limnol.*, 13, 195-211.

Likens, G. E. and Butler, T. J. (1981) Recent acidification of precipitation in North America, *Atmospheric Environment* (In Press).

Likens, G. E., Bormann, F. H., Johnson, N. M., Fisher, D. W. and Pierce, R. S. (1970) Effects of forest cutting and herbicide treatment on nutrient budgets in the Hubbard Brook watershed-ecosystem, *Ecol. Monogr.*, 40(1), 23-47.

Likens, G. E., Bormann, F. H. and Johnson, N. M. (1972) Acid rain, *Environment*, 14(2), 33-40.

Likens, G. E., Bormann, F. H., Pierce, R. S., Eaton, J. S. and Johnson, N. M. (1977) *Biogeochemisty of a Forested Ecosystem*, New York, Springer-Verlag.

Likens, G. E., Wright, R. F., Galloway, J. N. and Butler, T. J. (1979) Acid rain, *Scientific American*, 241(4), 43-51.

McElroy, M. B., Wofsy, S. C. and Yung, Y. L. (1977) The nitrogen cycle: perturbations due to man and their impact on atmospheric N_2O and O_3, *Phil. Trans. Roy. Soc.* (London) B, 277, 159-181.

Olson, J. S., Pfuderer, H. A. and Chan, Y. -H. (1978) Changes in the global carbon cycle and the biosphere, Oak Ridge National Laboratory, Environmental Sciences Division Publication No. 1050, ORNL/EIS-109.

O'Sullivan, P. E. (1979) The ecosystem-watershed concept in the environmental sciences – a review, *Internat. J. Environ. Studies*, 13, 273-281.

Redfield, A. C. (1958) The biological control of chemical factors in the environment, *Amer. Sci.*, 46, 205-221.

Redfield, A. C., Ketchum, B. H., and Richards, F. A. (1963) The influence of organisms on the composition of sea-water, in Hill, M. N. (ed.), *The Sea*, New York, Interscience, volume 2, 26-77.

Rühling, Å. and Tyler, G. (1973) Heavy metal pollution and decomposition of spruce needle litter, *Oikos*, 24, 402-416.

Shugart, H. H. and West, D. C. (1977) Development of an Appalachian deciduous forest succession model and its application to assessment of the impact of the chestnut blight, *J. Environ. Management*, 5, 161-179.

Siccama, T. G. and Smith, W. H. (1978) Lead accumulation in a northern hardwood forest, *Environ. Sci. and Tech.*, 12, 593-594.

Söderlund, R. (1977) NO_x pollutants and ammonia emissions — a mass balance for the atmosphere over Northwest Europe, *Ambio*, **6**(2-3), 118-122.

Svensson, B. H. and Söderlund, R. (eds), (1976) Nitrogen, phosphorus and sulphur — Global cycles, SCOPE Report 7, *Ecol. Bull. (Stockholm)*, **22**.

Waring, R. H. and Franklin, J. F. (1979) Evergreen coniferous forests of the Pacific Northwest, *Science*, **204**, 1380-1386.

Woodwell, G. M., Whittaker, R. H., Reiners, W. A., Likens, G. E., Delwiche, C. C. and Botkin, D. B. (1978) The biota and the world carbon budget, *Science*, **199**, 141-146.

Some Perspectives of the Major Biogeochemical Cycles
Edited by Gene E. Likens
© 1981 SCOPE

CHAPTER 7

Interrelationships Between the Cycles of Elements in Freshwater Ecosystems

D. W. SCHINDLER

Department of Fisheries and Oceans, Freshwater Institute,
Winnipeg, Canada

ABSTRACT

Interrelationships between the cycles of phosphorus, nitrogen, carbon, sulphur, and silicon are discussed, with the primary focus on the Experimental Lakes Area.

Phosphorus appears to be capable of controlling major parts of the cycles of the other four elements by increasing atmospheric inputs for phosphorus and nitrogen, by increasing sedimentation to the hypolimnion for silicon, and by stimulating sulphate reduction by increasing the amount of organic matter to decompose in the hypolimnion.

Sulphate reduction increases as sulphate concentrations are increased, causing increased generation of dissolved organic carbon by anoxic decomposition.

Nitrogen and carbon do not appear to affect phosphorus, sulphur, or silicon cycles directly.

7.1 INTRODUCTION

A detailed review of the relationships between several biogeochemical cycles in freshwater would be a nearly impossible task. For a paper of this length, it is necessary to severely restrict the number of cycles considered. Consequently, I shall treat only examples of the interrelationships between the cycles of phosphorus, nitrogen, carbon, sulphur, and silicon. All of these are elements essential for the nutrition of plants. The cycles of all have been disrupted as the result of man's pollution of the biosphere. In most cases, I shall use examples from my own experience to illustrate interrelations between the cycles. I have purposely done this in order to illustrate the complexity of interactions which can be stimulated by perturbations to one type of aquatic system.

7.2 THE EFFECTS OF PHOSPHORUS ON THE CYCLES OF OTHER ELEMENTS

For many years, phosphorus has been recognized as the nutrient which most frequently limits the standing crop and production of freshwater algae (reviewed by

Vollenweider, 1968). More recently, it has been discovered that the degree of eutrophication affecting lakes may usually be predicted quite accurately from phosphorus input, ignoring supplies of other nutritional elements (Vollenweider, 1976; Rast and Lee, 1978; Schindler et al., 1978). The fact that these models work so well suggests that other nutrients are never scarce enough to restrict the growth of all species of algae. For example, when ionic nitrogen is scarce, nitrogen-fixing blue-green algae often become dominant, and when silicon supplies are depleted, diatoms are replaced by non-silicious forms.

The utility of the above phosphorus models often appears to contradict predictions from laboratory bioassays, which frequently predict limitation by elements other than phosphorus. While such laboratory studies may indicate accurately which nutrient or nutrients are limiting at a particular moment, there is reason to believe that they are of little utility in guiding the management of eutrophication. Experiments done in small flasks, lasting only a few hours to a few weeks, simply do not account for the adaptability of an entire ecosystem. With its large diversity of latent species of organisms an ecosystem can respond to a wide diversity of conditions, and nutrient deficiencies can be corrected over a period of months or years (Schindler, 1977). There appears to be no reason to rely on a flask bioassay to provide predictive information about a whole aquatic ecosystem than there is to expect an LD-50 experiment on mice to guide our management of terrestrial ecosystems.

The numbers of each species of latent organism and the rates of dormant processes in lakes are usually so low as to defy measurement, so that the only way to assess the response of one chemical cycle in an ecosystem to changes in another cycle is to actually perturb the ecosystem. What follows below is based entirely on experimental alterations of nutrient cycles in whole lake systems.

7.3 THE EFFECT OF PHOSPHORUS ON THE NITROGEN CYCLE

When phosphorus 'input' is high with respect to nitrogen, the rate of growth or production of phytoplankton populations often becomes limited by nitrogen. When this happens, nitrogen-fixing Cyanophyceae usually outcompete other forms so that atmospheric nitrogen contributes to the nitrogen requirements of the plankton. In a recent literature review of aquatic nitrogen fixation, Flett et al. (1980) found that fixation became important when the N:P ratio in nutrient loading fell below 10:1 by weight. While production and growth of nitrogen-fixing bluegreen algae are often lower than for other forms, their colonies are typically quite large, and therefore less susceptible to grazing or other mortality factors than smaller forms (Schindler and Comita, 1972). As a result, under steady-state conditions, the total algal standing crop is usually comparable to that which develops when the supply of ionic nitrogen is large.

In three extreme cases, one from each of the temperate, subarctic, and arctic climatic zones, phosphorus alone was experimentally added to lakes. Even when natural sources of nitrogen were accounted for, the N:P input in each case was less than 3:1 by weight. In all three of these cases, nitrogen fixing algae appeared in the epilithiphyton rather than in the phytoplankton (Holloway, 1976; Persson et al., 1975; D.W. Schindler, unpublished data). While phytoplankton standing crop following such extreme treatment did not increase as rapidly as when both N and P were supplied, eventually it reached a magnitude comparable to that in lakes receiving both nutrients (Schindler, 1980, in press). In all three cases, the nitrogen content of the lakes increased, presumably due to fixation of atmospheric N_2, even though nitrogen input from other sources was unchanged.

Several whole-lake experiments in the Experimental Lakes Area* of northwestern Ontario were designed to yield information about the interplay between phosphorus and nitrogen. For example, Lake 227 (area 5.0 ha, mean depth 4.4 m), was fertilized for 6 years with an N:P ratio of 14:1 by weight. Algal standing crops were dominated by the green alga, *Scenedesmus* (Schindler et al., 1973), and no nitrogen fixation was detectable (Flett et al., 1980). In 1975, phosphorus was added to the lake as in previous years, but nitrogen additions were reduced to an N:P ratio of 5:1. A bloom of the bluegreen alga *Aphanizomenon gracile,* never previously recorded in the lake, appeared within weeks, and atmospheric nitrogen contributed 14 per cent to the lake's total nitrogen income in that year. Phytoplankton standing crop was lower in the first year after the change in nitrogen loading, but by the second year it was similar to that obtained with high N:P ratios, suggesting that a time lag of 1–2 years was necessary before a new steady-state was reached. Phytoplankton production was suppressed for two years, but by the third year had recovered to pre-1975 levels (Shearer and DeClecq, unpublished data). Despite the reduced application of fertilizer nitrogen after 1975, this element continued to increase in the lake (Figure 7.1).

The fact that at high N:P ratios bluegreen algae tend to be outcompeted by other species has been used advantageously to purposely reduce populations of bluegreen algae. In western Manitoba, rainbow trout (*Salmo gairdnerii*) have been cultured in shallow, hypereutrophic prairie ponds for many years. The ponds occupy an area of rich, easily leachable geological substrata, and are continuously fed by nutrient-rich groundwater (Barica, 1975). Trout culture had proved impossible in the most productive of these ponds because dense populations of bluegreen algae accumulated in surface waters, presumably because they could not be grazed by herbivores. Eventually, these bluegreen populations senesced and decayed, causing total anoxia in surface water and killing the trout before they reached harvestable size. Barica et al. (in press) added 7–14 g-N m^{-3} week^{-1} as ammonium nitrate to hypereutrophic prairie ponds and lakes where the accumulation and decomposition of the bluegreen alga *Aphanizomenon flos-aquae* had caused trout kills to occur

---

*More detail about the Experimental Lakes Area and its experiments may be found in several papers in *J. Fish. Res. Board Canada,* (1971), **28**(2) and (1973), **30**(10) and in *Canad. J. Fish. Aquat. Sci.* (1980), **37**(3).

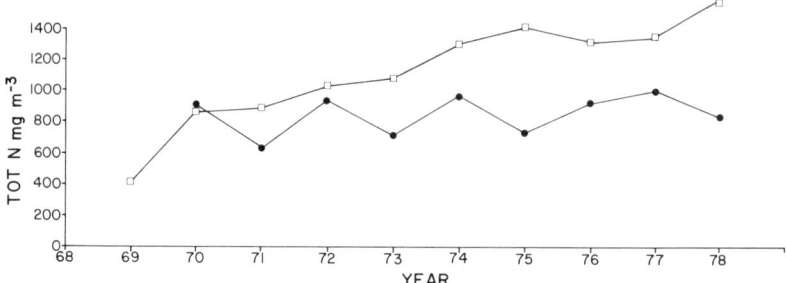

Figure 7.1 Total nitrogen concentration in Lake 227. The addition of nitrogen was cut to one third in 1975. □, whole lake annual average; ●, epilimnion, ice-free season average

almost every year. As a result of the nitrogen addition, small chlorophytes (*Scenedesmus* and *Oocystic* sp.) prospered, replacing bluegreens and eliminating the fish kill problem. Lower rates of addition caused reduction, but not complete elimination, of *Aphanizomenon* populations.

As a phosphorus-enriched lake becomes more eutrophic, the return of ammonia-nitrogen from sediments also increases relative to phosphorus. In Lake 227, ammonia released by epilimnetic sediments contributed 27 per cent of the total supply of nitrogen to the epilimnion of the lake (Schindler *et al.*, 1977). A further supply of nitrogen would be from hypolimnetic ammonia release, though this would only become available after overturn, in the spring and autumn of the year.

7.4 EFFECTS OF PHOSPHORUS ON THE CARBON CYCLE

After a lake is enriched with phosphorus, photosynthesis is usually stimulated, so that algae require more carbon to produce larger standing crops. In softwater lakes which have a low concentration of dissolved inorganic carbon (DIC), algae may consume a high proportion of the DIC content of the lake. Because 50 to 90 per cent of the DIC in such lakes is usually present as bicarbonate, an excess of hydroxyl ions is generated by photosynthesis:

$$HCO_3^- \xrightarrow{algae} CO_2 \to OH^-$$

As a result, the pH of such lakes usually increases. In extreme cases, pH values of nine or even ten may result (Schindler *et al.*, 1973). The gaseous CO_2 content of surface waters is depleted by photosynthesizing algae to many orders of magnitude below that of the overlying atmosphere. This enormous gradient encourages atmospheric CO_2 to enter the lake. The rate of entry is a function of turbulence and the degree to which exchange is enhanced by chemical conversion of invading CO_2 to bicarbonate, which is greater at higher pH (Emerson, 1975).

All species of photosynthetic algae help 'drive' this exchange process by maintaining the CO_2 gradient, in contrast to the requirement for 'specialist' bluegreen algae, which are required to fix atmospheric nitrogen.

For the first few years after enrichment with phosphorus and nitrogen, primary production in Lake 227 was limited for several hours each day by the supply of atmospheric CO_2 (Schindler and Fee, 1973). However, algal standing crops were maintained in proportion to total phosphorus concentrations. Maintenance of a high standing crop of algae with low productivity implies that under such conditions sinking, grazing or other mortality factors must be suppressed, as well as photosynthesis, when DIC is low. In Lake 227, the suppressed factor appears to have been the zooplankton population (D. Malley, personal communication). The major crustacean species in the lake declined during the first three years of fertilization, and populations remained low after that time. While the reason for the zooplankton decrease is not known with certainty, it is thought to be due to the inhibition of crustacean moulting which takes place at high pH (O'Brien and deNoyelles, 1972).

By 1974, after five years of enrichment, the DIC concentration in Lake 227 had increased to the point where photosynthesis was no longer carbon limited.

The development of a CO_2 gradient favouring invasion of carbon to lakes from the atmosphere is not confined to the Experimental Lakes. Calculating pCO_2 from pH, temperature and alkalinity or total CO_2, reveals that such gradients are common in eutrophic softwater lakes in summer (Schindler et al., 1975). Typically, pH values of nine or more for such lakes indicate that surface waters are depleted in gaseous CO_2 with respect to the atmosphere, so that invasion will take place.

The enhanced production of organic matter, which follows phosphorus enrichment, also causes changes in the decomposition of organic carbon. The degree and duration of hypolimnetic and sediment anoxia typically increase with accelerated eutrophication, so that methanogenic fermentation replaces oxic metabolism as the major decomposition pathway. Methane production becomes a significant part of the carbon cycle, at least when sulphate concentrations are low. The methane may be converted to CO_2 by oxidation, which is accomplished biologically (Rudd et al., 1974). If methane oxidation occurs under ice, anoxic conditions may result, causing suffocation of fishes and many other organisms.

I am left with the feeling that carbon is not the dynamic element which drives the entire aquatic food web, as envisioned only a decade ago. Instead, carbon appears to be a 'dutiful slave', which adjusts its cycle to enhance and support its 'master nutrient', phosphorus. This idea is not a new one (Redfield, 1958) but it appears worthy of further examination. A similar situation also may exist in terrestrial ecosystems (see chapter 6).

7.5 EFFECTS OF PHOSPHORUS ON THE SULPHUR CYCLE

Algae typically contain about the same proportions of phosphorus and sulphur, although concentrations of the latter element in the form of ionic sulphate are at

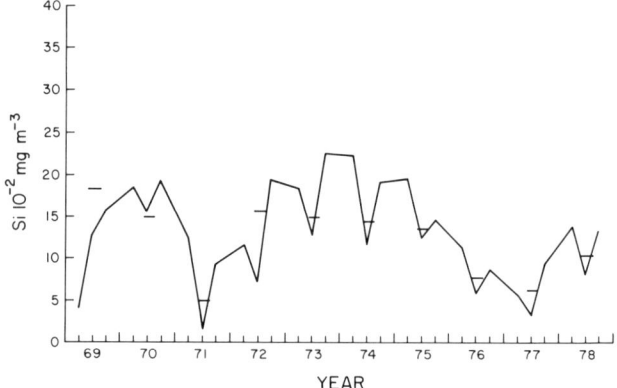

Figure 7.2 Reactive silicate concentrations in Lake 227. The line connects epilimnion values at spring overturn, midsummer low, and autumnal overturn. The horizontal lines are the mean epilimnion averages

least 1000 times more in freshwater. As a result, algal production has not been observed to cause significant depletion of epilimnetic sulphate.

However, important changes in the sulphur cycle may occur in the anoxic regions caused by phosphorus enrichment, as described above. Under anoxic conditions, sulphate is biologically reduced to sulphide. The oxygen stripped from the sulphate ion is used to drive the catabolism or organic matter by sulphate-reducing bacteria, which in turn stimulates the carbon cycle, as described below. Once in sulphide form, the sulphur may be precipitated as FeS in lakes where the concentration of iron is high, or accumulated in the hypolimnion as hydrogen sulphide when iron is low. Concentrations of hydrogen sulphide may be high enough to cause taste and odour problems in drinking water. Both of the above pathways depend on the pH of lake water as well as on iron and sulphide concentrations.

7.6 EFFECTS OF PHOSPHORUS ON THE SILICON CYCLE

The addition of phosphorus to lakes often causes depletion of the dissolved silica in surface waters. The silicon is sedimented with sinking diatoms into the hypolimnion and sediments, where most of it is released when the diatoms decompose. The major chemical effect of phosphorus is thus a relocation of silicon from epilimnion to hypolimnion during summer stagnation, followed by replenishment at overturn (Figure 7.2; Parker and Edgington, 1976; Conway et al., 1977).

As phosphorus inputs increase, the summer depletion of silica may be pronounced enough to limit the size of diatom populations. Typically, nitrogen is also in short supply, so that diatoms are replaced by bluegreen algae. The increasing replacement of diatoms in Lake Michigan by bluegreen algae during summer stagnation with increasing phosphorus input has been meticulously documented by

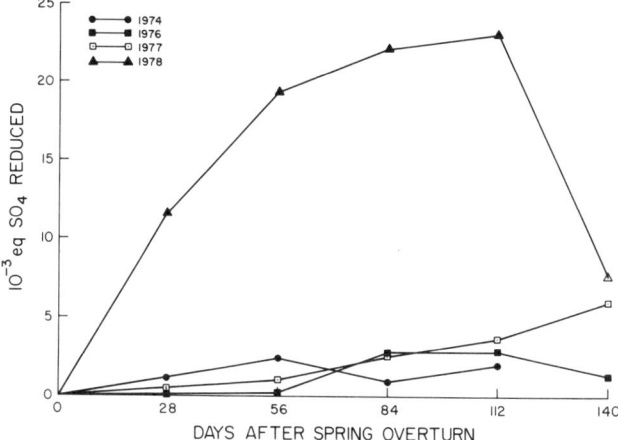

Figure 7.3 Sulphate reduction in the hypolimnion of Lake 223 before (1974) and during (1976–1978) acidification (from Schindler et al., 1980)

Schelske and Stoermer (1972). They hypothesize that increased burial of diatomaceous silicon which has resulted from the increased diatom production resulting from high phosphorus input in the twentieth century has left Lake Michigan with a chronic silica deficit. Unfortunately, due to the paucity of historical silica data for the Lake in winter, the magnitude of this deficit is not accurately known.

7.7 EFFECTS OF SULPHUR ON THE CARBON CYCLE

Recent studies have implicated sulphuric acid as the most abundant pollutant causing acid precipitation (e.g. Cogbill and Likens, 1974; see also chapters 3 and 4). Depletion of dissolved inorganic carbon reserves in softwater lakes occurs as hydrogen ions in precipitation are buffered by bicarbonate:

$$H^+ + HCO_3^- \rightarrow H_2CO_3 \xrightarrow{atm} H_2O + CO_2 \uparrow$$

As gaseous CO_2 generated by this mechanism exceeds the solubility of this gas in water, it is lost to the atmosphere. However, as described above, in anoxic hypolimnions or sediment waters, the sulphate increase which results from acidification may cause increased sulphate reduction. This reduction process appears to be limited by sulphate concentrations in softwater lakes. In a whole-lake experiment, sulphate concentrations were increased by 4X, all of it was reduced to sulphide in the course of a summer (Figure 7.3).

Sulphate reduction is thought to generate DIC according to the following reaction (Stumm and Morgan, 1970):

$$(CH_2O)_{106}(NH_3)_{16}H_3PO_4 + 53SO_4^{2-} \rightarrow 106DIC + 53S^{2-}$$
$$+ 16NH_3 + 106H_2O + H_3PO_4$$

The amount of carbon generated by this mechanism increases with the amount of available sulphate (Figure 7.3). Almost all of the sulphide produced by this reaction precipitates as FeS, so that hydrogen sulphide concentrations in the hypolimnion are usually undetectable (Schindler et al., 1980). Because iron concentrations in the hypolimnion of Lake 223 reach several tens of milligrams per litre when anoxia prevails, excess iron remains in the hypolimnion. In order to maintain charge balance, DIC in the hypolimnion becomes stored as bicarbonate at a rate which balances the permanent disappearance of iron sulphides (Schindler et al., 1980). Due to the slow reaction of even amorphous FeS (Berner, 1970), not all of the precipitated FeS is redissolved and reoxidized at spring and autumn turnover, i.e. the reaction is not completely reversible within the short period when oxic conditions prevail in the hypolimnion (R. B. Cook, unpublished data). As a result, the bicarbonate generated as described above becomes a permanent addition to the buffering capacity of the lake. The proportion of total buffering appears to increase as the sulphate concentration increases, which one would expect, as long as microbial sulphate reducers remain substrate limited and are not inhibited by high concentrations of hydrogen ion or toxic trace metals. Iron concentrations also must be high enough to precipitate sulphide efficiently. While a combination of such conditions may not be common in the hypolimnions of softwater lakes, the same processes may occur in sediments, which are typically anoxic below the top centimetre or so even in epilimnetic areas (Ben-Yaakov, 1973; Howarth and Teal, 1979). Prediction of the rate of reaction in such regions is hampered by our inadequate ability to predict diffusion of substances between sediments and overlying waters.

7.8 EFFECTS OF NITROGEN AND CARBON ON THE CYCLES OF OTHER ELEMENTS

Nitrogen and carbon were added to Lake 226SW. No artificial additions of phosphorus were made. No substantial changes in species, algal abundances or chemical processes were observed.

In a complimentary experiment, Lake 304 was rendered eutrophic by two years of fertilization with phosphorus, nitrogen, and carbon. In the third year, phosphorus additions were not made, but fertilization with nitrogen and carbon continued. Algal species and standing crops returned rapidly to values observed before fertilization.

Nitric acid has been implicated as constituting an average of 30 per cent of the strong acid in polluted rain (Cogbill and Likens, 1974; Galloway et al., 1976). As for sulphuric acid, a major impact on the carbon cycle would be expected, due to the effects of the hydrogen ion. If denitrified to dinitrogen, the nitrate ion could supply permanent buffering as described above for sulphate reduced to sulphide.

Such effects have not been studied. It appears that alteration of carbon and nitrogen inputs have little effect on the cycles of other elements. One indirect exception might be the addition of enough decomposable organic matter to a water body to generate anoxic conditions in the hypolimnion. Such conditions may occur when pulp and paper effluents or wastes from sugar processing are discharged into water. If the resulting anoxia causes sufficient decreases in redox potential at the mud-water interface, phosphorus which has been bound to ferric iron may become soluble as iron is reduced (Mortimer, 1941-42). This release does not always accompany anoxic conditions. In the Experimental Lakes, where phosphorus appears to be bound to organic materials in sediments (Jackson and Schindler, 1975), phosphorus is bound as strongly under anoxic as under oxic conditions (Schindler et al., 1980). A recent study of the mechanism of phosphorus release in Greifensee, Switzerland, revealed that diagenesis of phosphorus-bearing inorganic minerals took place under anoxic conditions (Imboden and Emerson, 1978), and it may be that lakes with primarily mineral sediments react differently from those where sediments have high concentrations of organic matter.

7.9 DISCUSSION

As we have seen, surprising and sometimes undesirable changes may occur in several aquatic chemical cycles due to perturbations of one of them. Often the perturbations are not even imposed directly upon the acquatic system. For example, alteration of the atmospheric cycles of sulphur and nitrogen have caused major changes in the aquatic carbon cycle via the acid rain phenomenon. Carbon originally stored in lakes as bicarbonate is transformed to CO_2 by reaction of acid with the bicarbonate, then lost to the atmosphere. Phosphorus and nitrogen added as fertilizers to terrestrial systems have made their way into aquatic systems, along with silicon eroded from tilled soils. Acidification of terrestrial soils by acid precipitation is thought to inhibit denitrification. This, plus the additional nitrate added with acidic rain, has caused greatly increased inputs of nitrate to receiving waters. Exchange of hydrogen ion from acid precipitation for aluminium in terrestrial soils has caused large quantities of the latter element to be discharged into lakes and streams, creating conditions lethal for fish (Cronan and Schofield, 1979; Baker and Schofield, 1980). The aluminium also appears to cause precipitation of phosphorus from lake water, possibly reducing the productivity of lakes (Dickson, 1980).

7.10 REFERENCES

Baker, J. P., and Schofield, C. L. (1980) Aluminium toxicity to fish as related to acid precipitation and Adirondack surface water quality, in D. Drabløs and A. Tollan (eds), *Ecological Impact of Acid Precipitation*, SNSF project, 1432 Ås-NLH, Norway, 292-293.

Barica, J. (1975) Collapses of algal blooms in prarie pothole lakes: their mechanism and ecological impact, *Verh. Internat. Verein. Limnol.*, **19**, 606-615.

Barica, J., Kling, H. and Gibson, J. (1980) Experimental manipulation of algal bloom composition by nitrogen addition. *Canad. J. Fish. Aquat. Sci.,* **37**(in press).
Ben-Yaakov, S. (1973) pH buffering of pore water of recent anoxic marine sediments, *Limnol. Oceanogr.,* **18**, 86-94.
Berner, R. A. (1970) Sedimentary pyrite formation, *Amer. J. Sci.,* **268**, 1-23.
Cogbill, C. V. and Likens, G. E. (1974) Acid precipitation in the northeastern United States, *Water Resour. Res.,* **10**(6), 1133-1137.
Conway, H. L., Parker, J. I., Yaguchi, E. M. and Mellinger, D. L. (1977) Biological utilization and regeneration of silica in Lake Michigan, *J. Fish. Res. Board Canad.,* **34**, 537-544.
Cronan, C. S. and Schofield, C. L. (1979) Aluminum leaching response to acid precipitation: effects on high-elevation watersheds in the Northeast, *Science,* **204**, 304-306.
Dickson, W. (1980) Properties of acidified water, in D. Drabløs and A. Tollan (eds) *Ecological Impact of Acid Precipitation,* SNSF Project 1432, Ås-NLH, Norway, 75-83.
Emerson, S. (1975) Chemically enhanced CO_2 gas exchange in a eutrophic lake: a general model, *Limnol. Oceanogr.,* **20**, 743-753.
Flett, R. J., Schindler, D. W., Hamilton, R. D. and Campbell, N. E. R. (1980) Nitrogen fixation in Canadian Precambrian Shield lakes, *Canad. J. Fish. Aquat. Sci.,* **37**, 494-505.
Galloway, J. N., Likens, G. E. and Edgerton, E. S. (1976) Acid precipitation in the northeastern United States: pH and acidity, *Science,* **194**, 722-724.
Holloway, H. (1976) Nitrogen fixation in the periphyton and phytoplankton of three lakes from the Experimental Lakes Area of northwestern Ontario, *B. Sci. Thesis,* McMaster University, Hamilton, Ontario.
Howarth, R. W. and Teal, J. M. (1979) Sulfate reduction in a New England salt marsh, *Limnol. Oceanogr.,* **24**, 999-1013.
Imboden, D. M. and Emerson, S. (1978) Natural radon and phosphorus as limnologic tracers: horizontal and vertical eddy diffusion in Greifensee, *Limnol. Oceanogr.,* **23**(1), 77-90.
Jackson, T. A. and Schindler, D. W. (1975) The biogeochemistry of phosphorus in an experimental lake environment: evidence for the formation of humic–metal–phosphate complexes, *Verh. Internat. Verein. Limnol.,* **19**, 211-221.
Mortimer, C. H. (1941-42) The exchange of dissolved substances between mud and water in lakes, *J. Ecol.,* **29**, 280-329; **30**, 147-201.
O'Brien, W. and deNoyelles, F., Jr. (1972) Photosynthetically elevated pH as a factor in zooplankton mortality in nutrient enriched ponds, *Ecology,* **53**(4), 605-614.
Parker, J. I. and Edgington, D. N. (1976) Concentration of diatom frustules in Lake Michigan sediment cores, *Limnol. Oceanogr.,* **21**, 887-893.
Persson, G., Holmgren, S. K., Jansson, M., Lundgren, A., Nyman, B., Solander, D. and Anell, C. (1975) Phosphorus and nitrogen and the regulation of lake ecosystems: Experimental approaches in subarctic Sweden, in *Proc. Circumpolar Conference on Northern Ecology Section 2,* September 1975, Ottawa, National Resource Council of Canada, 1-20.
Rast, W. and Lee, G. F. (1978) Summary analysis of the North American (US portion) OECD eutrophication project: nutrient loading–lake response relationships and trophic state indices, USEPA, EPA-600/3-78-008. Corvallis Experimental Research Laboratory, Corvallis, Oregon.
Redfield, A. C. (1958) The biological control of chemical factors in the environment, *Amer. Sci.,* **46**, 205-221.

Rudd, J. W. M., Hamilton, R. D. and Campbell, N. E. R. (1974) Measurement of microbial oxidation of methane in lake water, *Limnol. Oceanogr.*, **19**, 519-524.

Schelske, C. L. and Stoermer, E. F. (1972) Phosphorus, silica and eutrophication of Lake Michigan, in Likens, G. E. (ed.), *Nutrients and Eutrophication*, Spec. Symp. 1, Amer. Soc. Limnol. Oceanogr., 157-171.

Schindler, D. W. (1977) Evolution of phosphorus limitation in lakes, *Science*, **195**, 260-262.

Schindler, D. W. (1980) The effect of fertilization with phosphorus and nitrogen versus phosphorus alone on eutrophication of experiment lakes, *Limnol. Oceanogr.* (in press).

Schindler, D. W. and Comita, G. W. (1972) The dependence of primary production upon physical and chemical factors in a small, senescing lake, including the effects of complete winter oxygen depletion, *Arch. Hydrobiol.*, **69**, 413-451.

Schindler, D. W. and Fee, E. J. (1973) Diurnal variation of dissolved inorganic carbon and its use in estimating primary production and CO_2 invasion in Lake 227, *J. Fish. Res. Board Canad.*, **30**(10), 1501-1510.

Schindler, D. W., Fee, E. J. and Ruszczynski, T. (1978) Phosphorus input and its consequences for phytoplankton standing crop and production in the Experimental Lakes Area and in similar lakes, *J. Fish. Res. Board Canad.*, **35**(2), 190-196.

Schindler, D. W., Hesslein, R. and Kipphut, G. (1977) Interactions between sediments and overlying waters in an experimentally-eutrophied Precambrian Shield lake, in Golterman, H. L. (ed.) *Interactions Between Sediments and Freshwater*, Proc. Symp., Amsterdam, Sept. 1976; the Hague, Junk, and Wageningen, PUDOC, 235-243.

Schindler, D. W., Kling, H., Schmidt, R. V., Prokopowich, J., Frost, V. E., Reid, R. A. and Capel, M. (1973) Eutrophication of Lake 227, Experimental Lakes Area, northwestern Ontario by addition of phosphate and nitrate, Part 2, The second, third, and fourth years of enrichment, 1970, 1971 and 1972, *J. Fish. Res. Board Canad.*, **30**(10), 1415-1440.

Schindler, D. W., Lean, D. R. S. and Fee, E. J. (1975) Nutrient cycling in freshwater ecosystems, in *Productivity of World Ecosystems*, Proceedings of a symposium, Aug. 31-Sept. 1, 1972, Seattle, Washington, National Academy of Sciences, Washington, D.C., 96-105.

Schindler, D. W., Wagemann, R., Cook, R., Ruszczynski, T. and Prokopowich, J. (1980) Experimental acidification of Lake 223, Experimental Lakes Area. I. Background data and the first three years of acidification, *Canad. J. Fish. Aquat. Sci.*, **37**, 342-354.

Stumm, W. and Morgan, J. H. (1970) *Aquatic Chemistry: An Introduction Emphasizing Chemical Equilibria in Natural Waters*, New York, Wiley.

Vollenweider, R. A. (1968) The scientific basis of lake and stream eutrophication, with particular reference to phosphorus and nitrogen eutrophication factors, Paris, *Tech. Rep. OECD*, DAS/CSI/68, **27**, 1-182.

Vollenweider, R. A. (1976) Advances in defining critical loading levels for phosphorus in lake eutrophication, *Mem. Inst. Ital. Idrobiol.*, **33**, 53-83.

Some Perspectives of the Major Biogeochemical Cycles
Edited by Gene E. Likens
© 1981 SCOPE

CHAPTER 8

Interactions Between Major Biogeochemical Cycles in Marine Ecosystems

R. WOLLAST

University of Brussels, Belgium

ABSTRACT

The geochemical cycle of the elements in a marine system is strongly dependent on or influenced by biological activity. The production of organic matter acts as a major process of transfer of elements from the dissolved to the particulate state as soft or hard tissues. On the other hand, respiration or biological degradation may lead to the redissolution of particulate species.

As primary productivity is strictly limited to the photic zone and decay of organic matter is pursued in the deeper water masses of the oceanic system, the distribution of many elements exhibits a strong vertical gradient.

The biological activity also strongly affects master variables of seawater: pH, oxydo-reduction potential, and alkalinity. These changes may in turn induce chemical dissolution or precipitation reactions and influence the distribution of biologically inactive elements.

A general model of the cycle of various elements based on biological productivity and regeneration in a water column divided into a photic zone and a deep-water layer is presented here. It is based on the cycle of organic matter in the ocean acting as the driving force.

The vertical fluxes between the two water reservoirs are due to the settling of particulate organic and inorganic matter and to the transfer of dissolved species related to vertical advection and turbulent diffusion.

It is possible in this way, to build coherent global models describing the fluxes of organic carbon, phosphorus, nitrogen, silica and calcium carbonate in the ocean, and interrelations between the dynamics of the marine biota and the cycle of other minor elements may be expected.

8.1 INTRODUCTION

The first and necessary approach to modelling of chemical cycles in the ocean is to establish a mass balance model where the ocean is considered as a black box. It is generally assumed that a steady state is realized on the time scale considered, that is the composition of the system is independent of time. This model describes transport paths and fluxes among a limited number of physically well-defined portions of the earth like the continents, the atmosphere, the marine sediments.

The quality of such a model is controlled by the mass balance between the input and the output fluxes, which is equal to zero if no accumulation or removal in the reservoir is assumed (steady state). These models already permit an estimation of the importance of the modifications to natural chemical cycles owing to perturbations introduced by human activities. They are, however, insufficient to predict the effect of these perturbations, which requires furthermore that we also understand the mechanism responsible for the various fluxes involved in the model.

In the case of the geochemical cycles in the ocean, a first improvement was introduced by Sillen (1961) who considered that seawater could be viewed on a geologic time scale as an aqueous solution nearly in equilibrium with typical minerals (clays, carbonates, . . .). The thermodynamic model of seawater has serious limitations when short-term controls of chemical composition are considered. Also it is obvious that for many elements showing vertical concentration gradients in the ocean, which is the case for the elements vital to life, an equilibrium model is not applicable. Moreover, organic matter itself is thermodynamically unstable in the physicochemical conditions prevailing in the oceans.

The trend today is away from thermodynamic modelling and towards kinetic models, particularly those involving biological processes as kinetic controls on element cycles. Many elements are directly involved to a certain extent in biological processes, or their behaviour is indirectly affected by the physicochemical changes induced by the biological activity. Conversely, the biological activity itself is dependent on the available concentration of certain well defined chemical species in the marine environment, and the domain of potential activity of living organisms is in turn determined by thermodynamic factors. The interrelations between the cycles of the elements are for these reasons extremely intricate and difficult to handle without drastic simplifications.

I have chosen to present some tentative cycles for various elements based on the cycle of organic matter in the ocean considered as the driving force. The fluxes in the marine ecosystem of other elements vital to life will then be computed on the basis of the carbon cycle.

I shall first describe the model of organic carbon in the ocean as a driving parameter and show how the biological activity affects the distribution and behaviour of other elements. The relative importance of the biological processes in the cycle of these elements may further be compared with the global geochemical cycle restricted for the oceans to the continental input and the sedimentary output.

8.2 THE BIOGEOCHEMICAL CYCLE OF ORGANIC CARBON

The production of organic matter by photosynthesis is limited by the penetration of light and is thus restricted to the surface layer of the ocean (maximum depth of

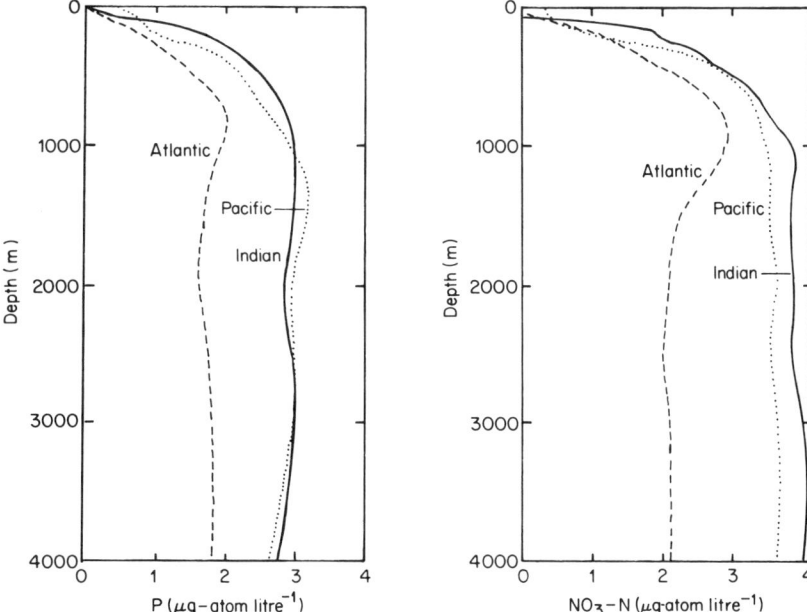

Figure 8.1 Mean vertical distribution of phosphate and nitrate in the oceans (from Svedrup et al., 1942)

200 metres). In addition to light, the rate of photosynthesis is also limited by the supply of nutrients, particularly nitrogen and phosphorus.

These nutrients are virtually exhausted (Figure 8.1) from the photic zone by incorporation in phytoplankton. The phytoplankton is consumed by marine animals and the excreta and dead bodies are bacterially degraded with regeneration of the nutrients either in the surface waters or in the deeper water masses, as the organic debris sinks. As a consequence the deeper waters are enriched in nutrients which may be restored to the photic zone by upwelling or by vertical turbulent mixing.

Finally, a fraction of the biogenic material reaches the sediment-water interface. Further degradation of the organic matter or dissolution of the skeleton proceeds until the reactive material has been either exhausted or covered by a sufficient layer of accumulating sediment.

A general model of the cycle of biological productivity and regeneration shows (Figure 8.2) a water column divided into a photic zone and a deep-water layer. The fluxes between these two reservoirs are due to the settling of particulate organic and inorganic matter from the surface waters to the intermediate and deep waters and to the transfer of matter related to advection and turbulent diffusion.

The model limited to the water column must also take into account the fluxes at the boundaries including river input, exchanges with the atmosphere, sedimen-

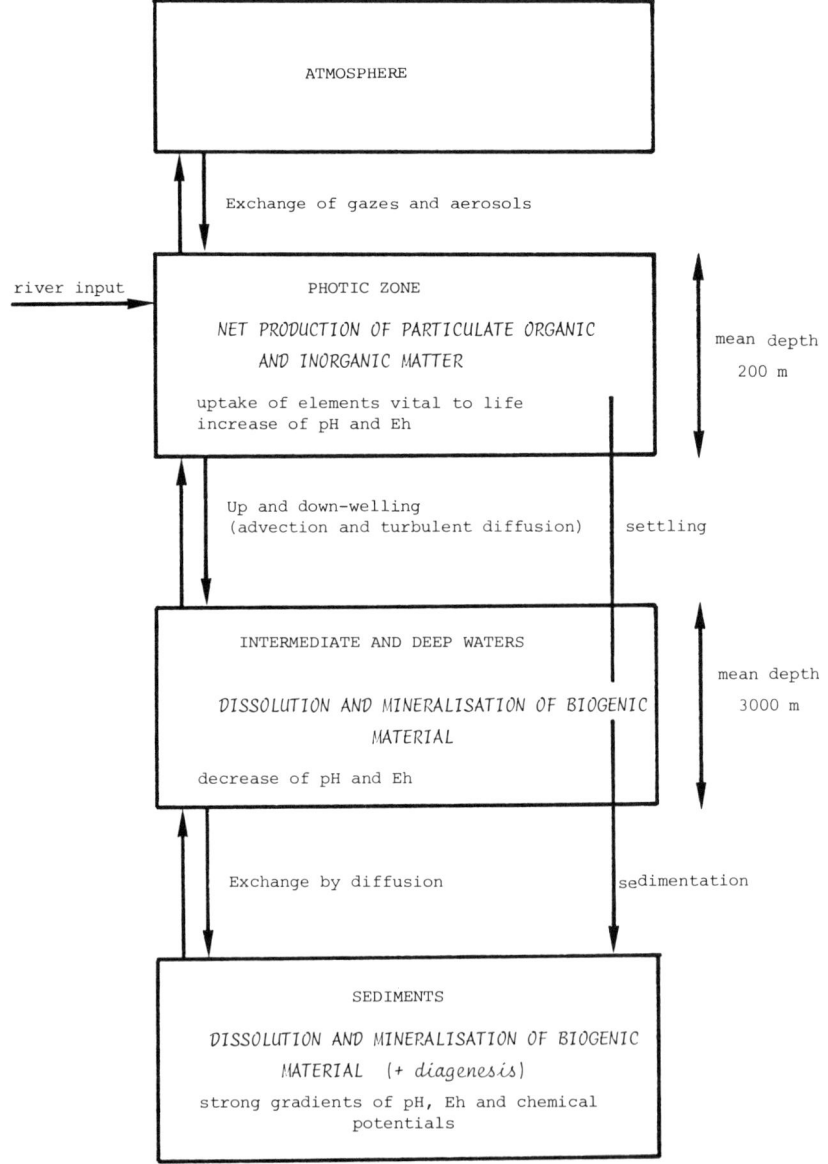

Figure 8.2 Major reservoirs and fluxes of the biogeochemical cycle of elements related to the biological activity in the ocean

tation of particulate matter at the bottom, and exchanges of dissolved species with the pore waters of the sediments by diffusion.

The transfer of compounds from the continents and from the atmosphere to the photic layer may be estimated from the global river discharge and the annual rainfall over the oceans multiplied by the mean concentration of the compounds in rivers and rain water. A total river discharge of 0.32×10^{20} g yr^{-1} and yearly precipitation on the oceans equal to 3.47×10^{20} g were used in the following calculations.

Besides the role played by the living organisms in the transfer of chemical species between reservoirs, it is also important to remember that the biological activity affects some master chemical variables such as the pH, the redox potential, and the total inorganic carbon content. Thus in the surface layer where photosynthesis exceeds respiration, the water exhibits a higher pH and a higher oxydoreduction potential whereas in the deeper water only respiration occurs which lowers the pH and the Eh.

The influence of photosynthesis and respiration on these two variables is well demonstrated if one considers the following reversible reaction:

$$CO_2 + 4H^+ + 4e^- \rightleftharpoons CH_2O + H_2O$$

Eventually the deep water may become anoxic if the flux of detrital organic matter and its subsequent decomposition exceeds the available quantity of dissolved oxygen. However, the anoxic conditions are mainly realized in the sediments where oxygen is generally depleted at a depth of a few centimetres.

These changes may indirectly affect the behaviour of some elements particularly those sensitive to changes of the oxydo-reduction potential like N, Fe, Mn, Cu, I, and so forth.

Quantitatively there are now rather coherent values for the carbon cycle in the ocean. de Vooys (in press) has critically reviewed the various recent estimates of the primary production in the marine environment and has concluded that 43.5×10^{15} g C yr^{-1} is the best approximation of primary production in seas and oceans now available.

The same author also collected some estimates from the literature concerning the per cent of total primary production which sinks to deeper levels and to the bottom. Some six per cent sinks from the photic zone, based mainly on the vertical distribution of particulate organic matter in the water column. A value of 0.6 per cent was chosen for the transfer of particulate organic matter to the sediments on the basis of the organic content of surficial sediments.

It must be pointed out that this organic matter is further subjected to biological degradation and according to Garrels and Perry (1974) only one third of the deposited organic material is finally preserved in the sediments.

There are quite large uncertainties concerning the concentration of particulate and dissolved organic matter in river and rain waters but their contribution is very small compared to the primary productivity.

I have adopted here the values of Garrels et al. for the river discharge which are respectively equal to 0.13×10^{15} g C yr^{-1} for dissolved organic carbon and 0.07×10^{15} g C yr^{-1} for particulate organic carbon.

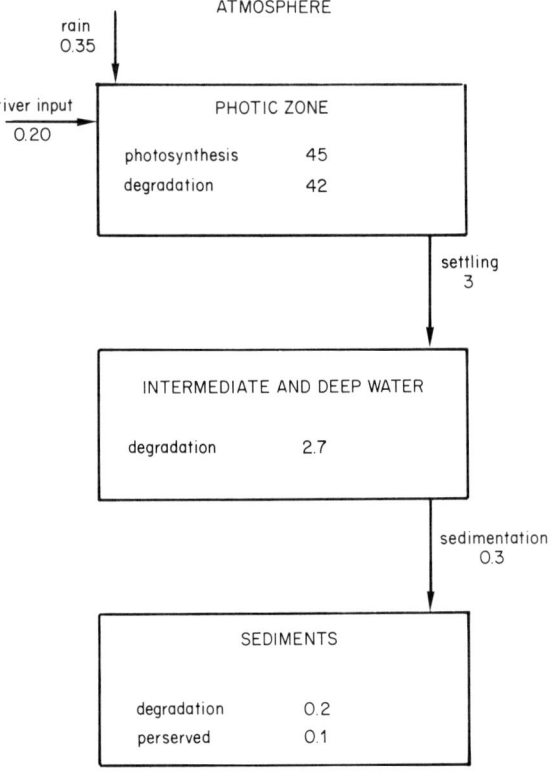

Figure 8.3 Tentative global cycle of organic matter in the ocean. Organic matter values times 10^{15} g C yr^{-1}

Maier and Swain (1978) have estimated from a range of data collected in the literature that a value of 1 mg C litre^{-1} in rain water may be used as representative of unpolluted atmospheric conditions. The corresponding flux over the ocean is then 0.35×10^{15} g C yr^{-1} for dissolved and particulate organic carbon.

A tentative global cycle of organic carbon in the ocean based on the data presented here is given in Figure 8.3. For convenience, the values used in this figure have been rounded but are well within the range of the existing uncertainties.

The biological activity characterized by the carbon cycle will be used in the next section as the driving force which controls the fluxes of various other biogenic elements in the oceanic system. The mean composition of the phytoplankton and of the suspended detrital organic matter in the various water layers and in the sedimentary column permits an evaluation of the fluxes of the particulate species between the reservoirs. The steady-state assumption applied to each reservoir then allows me to compute by difference the compensating fluxes of the dissolved species required to satisfy the mass balance.

I shall discuss successively the case of the classical nutrients N and P, of the biogenic portions of SiO_2 and $CaCO_3$ and finally of some trace elements vital to life.

8.3 THE PHOSPHORUS AND NITROGEN CYCLES

I have adopted the well-known Redfield ratio 106C:16N:1P (Redfield et al., 1963) for the mean composition of the marine plankton. This ratio allows me to compute from the carbon cycle the amount of nutrients consumed and mineralized in the photic zone and the amount of particulate P and N transferred to the deep waters as organic detrital matter.

The mineralization of the organic N and P compounds in the intermediate and deep water has been slightly increased with respect to the organic carbon. This increase is justified by the fact that many N and P compounds are more rapidly degraded and that, as a consequence, the C:P and C:N ratios in the suspended matter tend to increase with depth in the water column.

8.3.1 The Phosphorus Cycle

The results for the phosphorus cycle are presented in Figure 8.4. The amount of phosphorus preserved in the sediments has been taken from Lerman et al. (1975) and is based on the mean composition of marine sediments. It should be noted that these authors have computed a more detailed model including the terrestrial cycle. The fluxes calculated here are in good agreement with their previous estimations, shown in parentheses in Figure 8.4.

The validity of this model was tested further from independent evaluations of the fluxes of nutrients between the intermediate water and the photic zone due to upwelling and turbulent diffusion.

The rate of vertical mixing in the ocean has been estimated mainly from the vertical distribution of natural isotopes like ^{14}C, ^{32}Si, and ^{226}Ra or man-made isotopes like ^{3}H, ^{90}Sr, ^{137}Cs (Broecker, 1974).

A rather consistent mean of 3 ± 1 m yr^{-1} may be deduced from the collected data (Lerman, 1979) for the rate of mixing of deep and intermediate water with surface water. If this value is multiplied by the mean concentration of dissolved phosphate in the deep water (70×10^{-3} g P m^{-3}) taken from Figure 8.1, I obtain a vertical flux of 75×10^{12} g P yr^{-1}. This value is in remarkable agreement with the evaluation based on the carbon cycle discussed above (71×10^{12} g P yr^{-1}).

It is also possible to evaluate the relative importance of upwelling and turbulent diffusion to the vertical mixing. The turbulent diffusion across the thermocline is a very slow process which may be considered as a limiting step. Lerman (1979) has reviewed the available data for the oceans and lakes and from his work I have selected a mean value of 1500 m^2 yr^{-1} (0.5 cm^2 s^{-1}) for the eddy diffusion coefficient.

From Figure 8.1, the mean concentration gradient of dissolved phosphate is approximately 60×10^{-6} g P m^{-3} m^{-1} which gives over the total area of the ocean a vertical flux by diffusion of 35×10^{12} g P yr^{-1}.

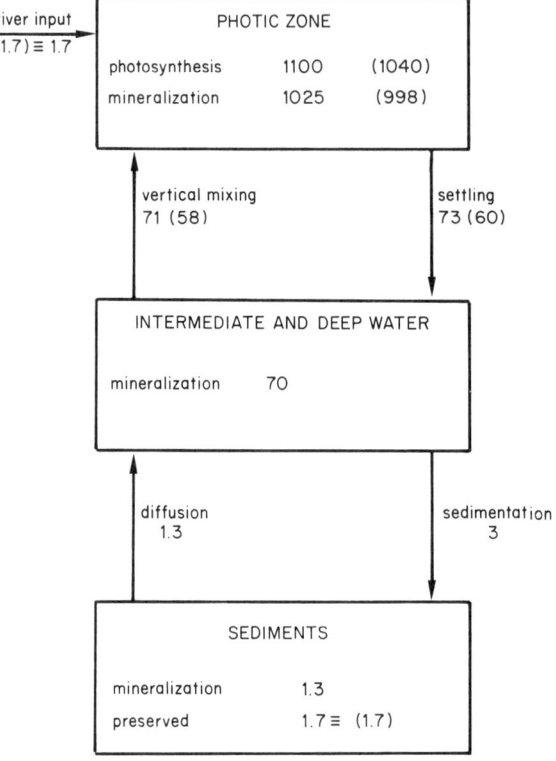

Figure 8.4 Tentative cycle of phosphorus in the ocean. The figures in brackets are independent estimates from Lerman et al. (1975) ($\equiv$) indicates values adopted here from these authors. Phosphorus values times 10^{12} g P yr^{-1}

Thus approximately one half of the vertical flux of phosphate is due to turbulent diffusion and the other half to advection (upwelling). However, upwelling is mainly restricted to coastal zones whereas turbulent diffusion prevails in the open ocean. The differences in the surface areas of these two zones explain well the drastic differences of primary productivity observed between coastal regions and the open sea.

8.3.2 The Nitrogen Cycle

The nitrogen cycle is more complicated due to the various species of inorganic N (NH_4^+, N_2, N_2O, NO_2^-, NO_3^-) involved in the biological cycle and to the influence of the local oxydo-reduction conditions on the thermodynamic stability of these species. The energetic implications related to the thermodynamic properties of the nitrogen species control in turn the possible pathways of the bacteriological activity.

Interactions Between Major Biogeochemical Cycles in Marine Ecosystems 133

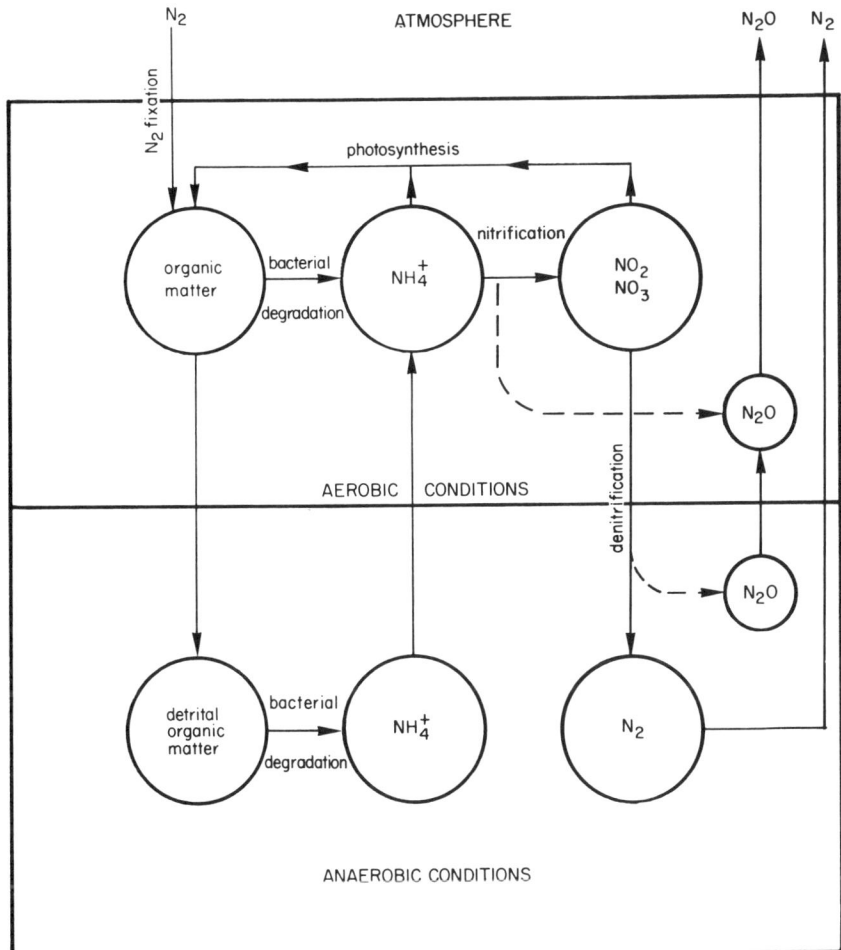

Figure 8.5 Schematic biological pathways of nitrogen in aerobic and anaerobic environments

I have considered here a rather simplified scheme (Figure 8.5) where a distinction is made between aerobic and anaerobic environments.

A tentative cycle for nitrogen is presented in Figure 8.6. The oceanic waters are generally sufficiently oxygenated that nitrate is usually the final more stable compound in solution. I do not have sufficient data on the restricted cases of anaerobic conditions to evaluate the amount of denitrification which may occur in the intermediate and deep waters. On the other hand, oxygen is rapidly depleted in the pore water of marine sediments and the anaerobic conditions prevailing there are favourable to the denitrification process. The tentative figures given for the cycle of nitrogen in the sediments are based on an extended study of North Sea sediments (Billen, 1978).

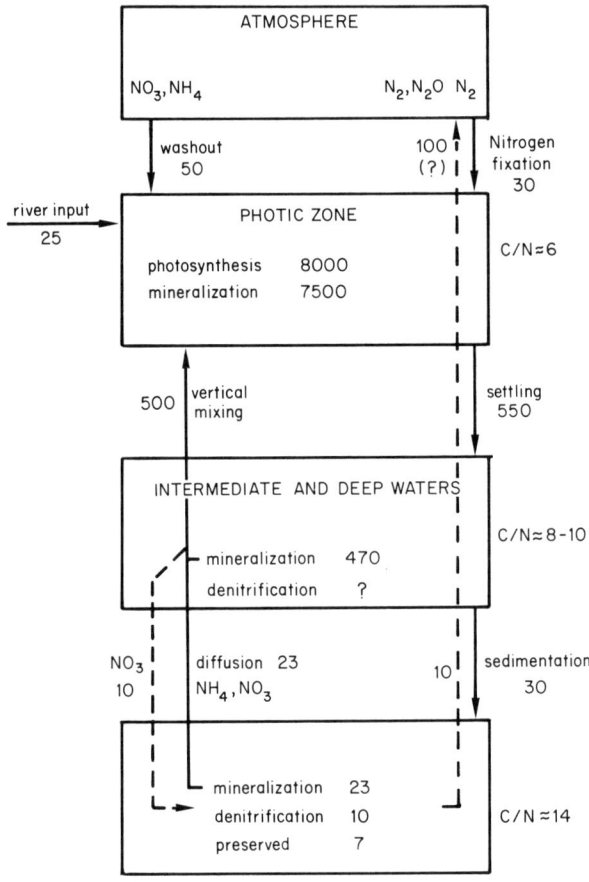

Figure 8.6 Tentative cycle of nitrogen in the ocean. Nitrogen values times 10^{12} g N yr^{-1}

According to Söderlund and Svensson (1976) the total nitrogen flux due to river runoff to the ocean is between 13 and 24 × 10^{12} g N yr^{-1}. However, more recent estimates are slightly higher: 20 × 10^{12} g N yr^{-1} (McElroy, 1976), 30 × 10^{12} g N yr^{-1} (Sweeney et al., 1977) and 35 × 10^{12} g N yr^{-1} (Delwiche and Likens, 1977). An intermediate value of 25 × 10^{12} g N yr^{-1} is adopted here.

For the rainout, fallout, and washout of nitrogen species over the ocean, again the range of the various estimates is broad: 30 to 80 × 10^{12} g N yr^{-1} (Söderlund and Svensson, 1976) and a mean value of 50 × 10^{12} g N yr^{-1} has been selected here.

In the photic zone, upwelling and vertical mixing remain the dominant sources of nitrogen but the river input and the atmospheric washout are relatively more important sources than in the case of phosphorus.

The extent to which nitrogen fixation occurs in the marine system is not accurately known and was estimated to be 20-120 × 10^{12} g N yr^{-1} (Söderlund and Svensson, 1976). I have selected here an estimate of 30 × 10^{12} g N yr^{-1}, as the best approximation.

It should be noted that the total mass balance for nitrogen in the system would require that approximately 100 × 10^{12} g N yr^{-1} are restored to the atmosphere as N_2 and N_2O). Here again there are not sufficient data about this flux and my value is subject to great uncertainties. The total nitrogen flux due to oceanic denitrification was estimated to vary between 25 and 179 × 10^{12} g N yr^{-1} by Söderlund and Svensson (1976).

As for phosphorus, it is possible to test the validity of the vertical flux due to vertical mixing. If I apply the same values as previously for the vertical mixing to the mean dissolved nitrogen concentration of the deep waters I obtain a vertical flux of 450 × 10^{12} g N yr^{-1}, which is again in good agreement with the computed value from the carbon cycle (500 × 10^{12} g N yr^{-1}).

The contribution of vertical turbulent diffusion to this flux, calculated on the basis of a mean concentration gradient of dissolved nitrogen of 280 × 10^{-6} g N m^{-3} m^{-1}, is equal to 160 × 10^{12} g N yr^{-1}. The relative importance of advection and turbulent diffusion is very similar to what I have observed for phosphorus and the same conclusions as earlier may be drawn here concerning the distribution of the productivity in the oceans.

8.4 THE CYCLES OF OPAL AND CALCIUM CARBONATE

The skeletons of the organisms produced in the photic zone represent an important weight fraction of the biogenic material. They are mainly composed of opal, an amorphous variety of silica and calcium carbonate, either as calcite or aragonite. After death of the organisms, the organic coating of the skeleton is rapidly removed and the mineral phase may dissolve during sinking if the water masses are undersaturated with respect to this phase. The cycles of silica and calcium carbonate are thus also intimately related to biological activity.

8.4.1 The Silica Cycle

More than 80 per cent of the silica is consumed by photoactive organisms (diatoms and silicoflagellates) in the surface waters. It is thus possible to connect the rate of uptake of dissolved silica directly to the primary productivity if the ratio of amorphous silica to organic carbon is known for the plankton.

There are large discrepancies for this ratio in the literature ranging from 2.3 for pure diatom blooms (Lisitzin, 1972) to 0.35 for a mean composition of the marine phytoplankton (Martin and Knauer, 1973). Considering the seasonal and geographical variations in the activity of the diatoms, I have selected a mean ratio of 0.6.

The annual production of opal in the photic zone is then estimated to be 25×10^{15} g SiO_2 yr^{-1}.

Seawater is always strongly undersaturated with respect to opal which constitutes the skeleton of these organisms and their frustules are thus submitted to rapid dissolution, especially in the surface waters where dissolved silica is virtually exhausted. Only three per cent of the biogenic opal reaches the sediment where dissolution continues. Finally only 1.5 per cent of the initial production is accumulated in the sedimentary column, mainly combined with clay minerals (Wollast, 1974; Berger, 1976). This accumulation is, however, balanced exactly by the river input.

The various steps of the behaviour of silica have been discussed previously and remain valid (Wollast, 1974) but the cycle presented here has been slightly modified to take into account more recent values for the vertical mixing in the oceans. The upward flux of dissolved silica, for a vertical mixing rate of 3 m yr^{-1} and a mean concentration of 8.4 g SiO_2 m^{-3} is equal to 9.1×10^{15} g SiO_2 yr^{-1}.

From the mass balance, it may be assumed that 60 per cent of the opal skeletons are thus redissolved in the photic layer. This value is well within the range of the rapid decrease of siliceous frustules with depth observed within the ocean (Kozlova, 1964; Lisitzin, 1972; Berger, 1976).

This cycle presented in Figure 8.7 shows again the considerable importance of the biological activity on the dynamics controlling the behaviour of a typical mineral compound in the ocean.

8.4.2 The Calcium Carbonate Cycle

Calcium carbonate is mainly produced in the water column by coccolithophores, foraminifera (both as calcite) and pteropods (as aragonite). The ratio method used earlier for silica is not applicable to carbonates because of the abundance of foraminifera and pteropods which are not primary producers. However, it is still possible to find some mean empirical factor to relate carbonate fixation to primary production (Berger, 1976).

Broecker (1974) estimates that particulate matter falling from the surface into the deep sea contains approximately 0.5 moles of silica and 0.5 moles of calcium carbonate per mole of organic matter. In more recent measurements of the composition of the particulate matter in the surface waters of the ocean, Lal (1977) found a mean weight content of 60 per cent of organic matter. Although considerable geographical differences occur, the mineral phase (40 per cent) contains on the average equal amounts of silica and calcium carbonate.

In a production-regeneration model in oceans based on the vertical fluxes in the water column and mass balances for the various reservoirs, Lerman (1979) has evaluated the fluxes of organic carbon, silica, and calcium carbon that leave a one kilometre thick surface layer, and found respectively 3.5×10^{15} g C yr^{-1}, (or 8.8×10^{15} as CH_2O yr^{-1}), 10×10^{15} g SiO_2 yr^{-1}, and 12.2×10^{15} g $CaCO_3$ yr^{-1}.

Thus I conclude that the flux of $CaCO_3$ leaving the photic zone must be nearly

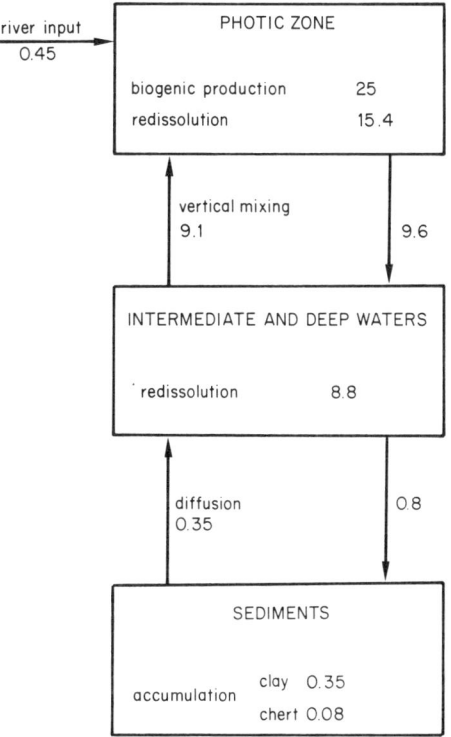

Figure 8.7 Tentative cycle of silica in the ocean. Silica values times 10^{15} g SiO_2 yr^{-1}

equal to the flux of organic matter (7.5×10^{15} g yr^{-1}) and of opal (9.6×10^{15} g SiO_2 yr^{-1}) computed in our previous cycles.

The surface waters are always oversaturated with respect to calcite and aragonite and thus the carbonate skeletons should be entirely preserved in the photic zone. It then may be concluded that the biological production of $CaCO_3$ which is equal to the flux of particulate inorganic carbonate would probably be close to 8.5×10^{15} g $CaCO_3$ yr^{-1}.

From the vertical distribution of calcareous shells, Berger (1976) concluded that only about one sixth of the biological production is finally preserved in the sediments. In fact, the water of the ocean becomes with depth, successively undersaturated with respect to aragonite and calcite. Further dissolution occurs in the sediments during early diagenesis. The amount of calcium carbonate accumulated yearly in the sedimentary record is rather well documented and may be estimated at 1.5×10^{15} g $CaCO_3$ yr^{-1} (Turekian, 1965; Garrels and Mackenzie, 1972).

Taking into account the preservation ratio of Berger cited above, an initial production of 9×10^{15} g $CaCO_3$ yr^{-1} may be calculated which is in good agreement

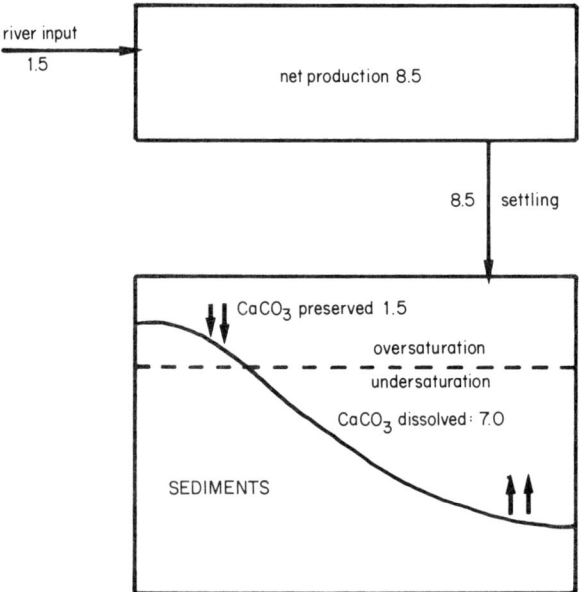

Figure 8.8 Tentative cycle of calcium carbonate in the ocean. Calcium carbonate values times 10^{15} g $CaCO_3$ yr^{-1}

with the previous estimate. The tentative cycle for calcium carbonate is given in Figure 8.8.

8.5 THE CYCLE OF MINOR ELEMENTS

Besides the major nutrients N and P, and the skeleton materials SiO_2 and $CaCO_3$, several minor elements are vital to life and their cycle may be controlled in the same way, that is by the biological activity in the ocean. Many trace elements exhibit a minimum of their concentration in the surface waters which corresponds to the hypothesis of a depletion in the surface waters by biological uptake and an increase of concentration in the intermediate and the deep water due to remineralization and dissolution of the biogenic material. However, a similar distribution may be obtained in some cases if the main source of one element for the open ocean is constituted by the atmospheric input of particulate matter which is submitted to dissolution during settling.

It is possible to obtain a better proof of the validity of this hypothesis by checking the eventual correlations existing between the vertical distribution of the element considered and the concentration of major nutrients like N, P, SiO_2. At the present time, the data concerning the mean elemental composition of the marine plankton are extremely scattered and vary often by several orders of magnitude partly due to the difficulties encountered in isolating the planktonic material in

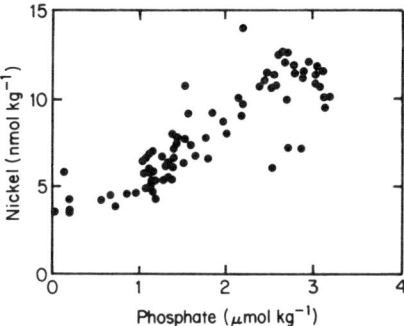

Figure 8.9 Relation between dissolved nickel and phosphate in the ocean (from Sclater et al., 1976)

the sea without contamination, and also probably to large geographical and climatological changes. It is thus difficult to evaluate quantitatively the importance of the biological activity on the cycle of minor elements in the ocean.

To demonstrate the possible importance of the biological processes on the distribution of trace elements in the oceans, I have selected four examples where the concentration of the element is strongly depleted in the surface waters and where there is a strong correlation with the concentration of another element whose distribution is known to be controlled by biological activity.

The first case is related to the distribution of nickel in the Atlantic and Pacific. Sclater et al. (1975) showed that the vertical distribution of dissolved nickel is actively controlled by the biological cycle and is closely related to the distribution of both phosphate and silicate (Figure 8.9).

Boyle et al. (1976) have demonstrated that cadmium profiles in the Pacific resemble those of phosphate and that the covariance with phosphate (Figure 8.10) suggests that cadmium is regenerated in shallow waters like the labile nutrients, rather than deeper in the ocean as for silicates.

On the other hand Bruland et al. (1978) reported a stronger correlation between zinc and silicon (Figure 8.11) than between zinc and phosphorus and nitrogen. The marked surface depletion of Zn and the fact that phytoplankton concentrates large amounts of this element all suggest that diatoms play an important role in the biogeochemical cycling of this element.

A similar correlation was found for aluminium (Figure 8.12) in the Mediterranean by Mackenzie et al. (1978) but a more critical analysis of other similar results indicates that aluminium is probably not only consumed by diatoms but also by a larger spectrum of the plankton (Caschetto and Wollast, 1979).

8.6 CONCLUSIONS

I have shown that it is possible to build coherent global models describing the fluxes of organic carbon, nitrogen, phosphorus, silica, and calcium carbonate in

140 Some Perspectives of the Major Biogeochemical Cycles

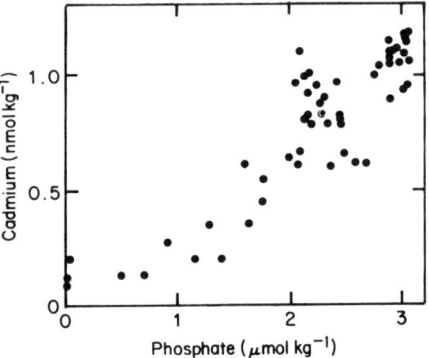

Figure 8.10 Cadmium concentration against phosphate for three Pacific profiles (from Boyle et al., 1976)

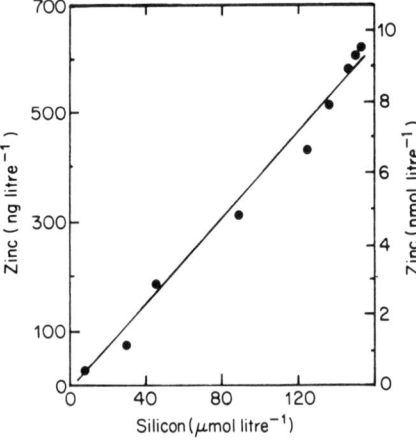

Figure 8.11 Zinc versus dissolved silica in the northeast Pacific (from Bruland et al., 1978)

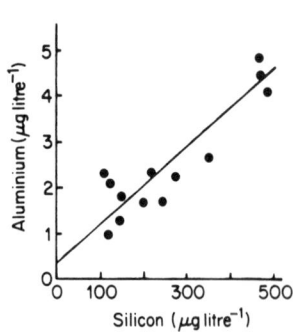

Figure 8.12 Covariation of dissolved aluminium and silicon (from Mackenzie et al., 1978)

the ocean, by considering the existing interrelations between these compounds in the dynamics of the marine biota. It may be expected that similar interrelations apply for many trace elements vital to life.

The study of these cycles is not only important for a better understanding of the phenomena controlling the various steps of the biological activity of marine organisms. It is becoming more and more evident that these biological processes are important kinetic controls on the concentration and distribution of elements within the ocean.

8.7 REFERENCES

Berger, W. H. (1976) Biogenous deep sea sediments, in Riley, J. P. and Chester, (eds), *Chemical Oceanography*, London, Academic Press, volume 5, 265-388.
Billen, G. (1978) A budget of nitrogen recycling in North Sea sediments of the Belgian coast, *Estuarine and Coastal Marine Science*, 7, 127-146.
Boyle, E. A., Sclater, F. and Edmond, J. D. (1976) On the marine geochemistry of cadmium, *Nature*, 263, 42-44.
Broecker, W. S. (1974) *Chemical Oceanography*, New York, Harcourt Brace Jovanovich.
Bruland, K. W., Knauer, G. A. and Martin, J. H. (1978) Zinc in the north-east Pacific water, *Nature*, 271, 741-743.
Caschetto, S. and Wollast, R. (1979) Vertical distribution of dissolved aluminum in the Mediterranean sea, *Marine Chemsitry*, 7, 151-155.
Delwiche, C. C. and Likens, G. E. (1977) Biological response to fossil fuel, in Stumm, W. (ed.), *Global Chemical Cycles and their Alteration by Man*, Berlin, Dalhem Conference, 73-88.
de Vooys, C. G. N. (1979) Primary production in aquatic environments, in *SCOPE Workshop on Biogeochemical Cycling of Carbon*, in press.
Garrels, R. M. and Mackenzie, F. T. (1972) A quantitative model for the sedimentary rock cycle, *Marine Chemistry*, 1, 27-41.
Garrels, R. M. and Perry, E. A., Jr. (1974) Cycling of carbon, sulfur and oxygen through geologic time, in Goldberg, E. D. (ed.), *The Sea*, New York, Wiley, volume 5, 303-306.
Kozlova, O. G. (1964) *Diatoms of the Indian and Pacific Sectors of the Pacific*, Moscow, Akad. Nauk.
Lal, D. (1977) The oceanic microcosm of particles, *Science*, 198, 997-1009.
Lerman, A., Mackenzie, F. T. and Garrels, R. M. (1975) Modelling of geochemical cycles: Phosphorus as an example, *Geol. Soc. Amer. Mem.*, 142, 205-218.
Lerman, A. (1979) *Geochemical Processes: Water and Sediment Environments*, New York, Wiley.
Lisitzin, A. P. (1972) *Sedimentation in the World Ocean*, Society of Economic Paleontologists and Mineralogists, Special Publication No. 17.
Mackenzie, F. T., Stoffyn, M. and Wollast, R. (1978) Aluminium in sea-water: control by biological activity, *Science*, 199, 680-682.
Maier, W. J. and Swain, W. R. (1978) Lake Superior organic carbon budget, *Water Res.*, 12, 403-412.
Martin, J. H. and Knauer, G. A. (1973) The elemental composition of plankton, *Geochim. Cosmochim. Acta*, 37, 1639-1653.
McElroy, M. B. (1976) Chemical process in the solar system: a kinetic perspective, in Herschbach, D. (ed.) *MTP International Review of Science*, London, Butterworth, 127-211.
Redfield, A. C., Ketchum, B. H. and Richards, F. A. (1963) The influence of organisms on the composition of sea-water, in M. N. Hill (ed.), *The Sea*, New York, Wiley-Interscience, volume 2, 26-77.
Sclater, F. R., Boyle, E. and Edmond, J. M. (1976) On the marine geochemistry of nickel, *Earth and Planetary Sci. Letters*, 31, 119-128.
Sillen, L. G. (1961) The physical chemistry of sea-water, in Sears, M. (ed.), *Oceanography*, Washington, D.C., AAAS, 549-581.
Söderlund, R. and Svensson, B. H. (1976) The global nitrogen cycle, in Svensson, B. H. and Söderlund, R. (eds), *Nitrogen, Phosphorus and Sulphur—Global Cycles*, SCOPE Report 7, *Ecol. Bull. (Stockholm)*, 22.

Svedrup, H. U., Johnson, M. W. and Fleming, R. H. (1942) *The Oceans*, New York, Prentice Hall.

Sweeney, R. E., Liu, K. K. and Kaplan, I. R. (1977) Oceanic nitrogen isotopes and their uses in determining the source of sedimentary nitrogen, in *Proc. International Symposium on Stable Isotope Chemistry*, New Zealand.

Turekian, K. K. (1965) Some aspects of the geochemistry of marine sediments, in Riley, J. P. and Skirrow (eds), *Chemical Oceanography*, New York, Academic Press, volume 2, 81–126.

Wollast, R. (1974) The silica problem, in Goldberg, E. D. (ed.), *The Sea*, New York, Wiley-Interscience, volume 5, 359–392.

SECTION III

Socio-Economic Impacts on Biogeochemical Cycles

Some Perspectives of the Major Biogeochemical Cycles
Edited by Gene E. Likens
© 1981 SCOPE

CHAPTER 9

Socio-Economic Impacts of the Effects of Man on Biogeochemical Cycles: Sulphur

G. PERSSON

Swedish Environment Protection Board, Solna, Sweden

ABSTRACT

Acidification of poorly buffered freshwaters accompanied by losses in fish population has occurred in large regions of northern Europe and eastern North America. These regions receive acid precipitation (pH 4.5) and are underlain by granitic or similar bedrock with thin and patchy soils.

Socio-economic effects of damage to the freshwater aquatic system are losses in commercial fishing and income from tourism including sport fishing. Such effects may be expressed in monetary terms. This is not the case for the destruction of freshwater ecosystems as such. An alternative evaluation would be to calculate the cost for restoring damaged ecosystems. The costs of liming are estimated.

There are a number of technical solutions available to reduce sulphur dioxide emissions. The obstacles are mainly of an economic nature. The typical costs of fuel oil and flue gas desulphurization are given.

Great variations exist in emission density and in emission per capita of sulphur dioxide within the European region. A reduction in sulphur dioxide emissions in Europe from the present level of 60 million tons to 25 million tons per year would cost about $10 billion ($$10^{10}$) annually.

9.1 INTRODUCTION

Atmospheric studies in recent years have substantially extended our knowledge of the origin and behaviour of sulphur oxides in the atmosphere. It has been shown that sulphur oxides travel over long distances and emissions in one country can contribute significantly to sulphate aerosol concentration and acid deposition in another country (OECD, 1977).

Sulphate aerosols have an indisputably adverse effect on visibility. Among obvious types of damage associated with visibility impairment are aesthetic and psychological costs, loss of property values, loss of tourist revenues in scenic areas, reduction in sunlight, hindrance to aviation, and general citizen dissatisfaction. Sulphate

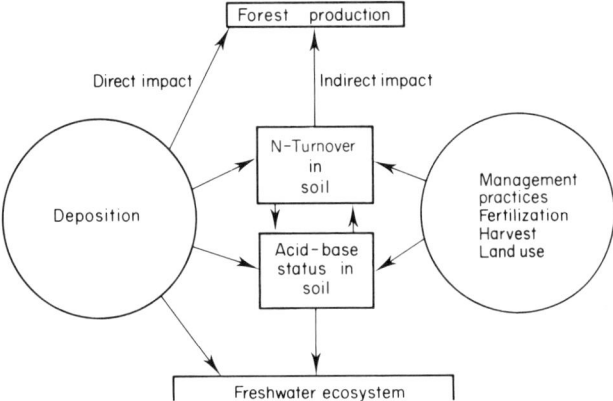

Figure 9.1 Acid deposition affects freshwater and forest ecosystems directly and indirectly (by changing the nitrogen turn-over and acid-base status in soil). The indirect impact is difficult to separate from the impact of local changes in land use and management

aerosols may also pay a significant role in climate modification as important constituents in atmospheric fine particulates affecting the transfer of radiative energy in the atmosphere.

Suspected health effects of sulphate aerosols are not fully supported by existing data.

Wet and dry deposition of acids, due primarily to anthropogenic sulphur dioxide emissions, is now recognized as a threat to terrestrial and aquatic ecosystems on both sides of the North Atlantic.

This paper deals with the socio-economic impacts of the ecological effects of acid deposition, and the methods and costs for sulphur dioxide abatement.

9.2 ECOLOGICAL EFFECTS OF ACID DEPOSITION

There is no simple relationship between acid deposition and effects on freshwater and terrestrial ecosystems (Figure 9.1). The effect of acid deposition on lakes and streams depends on the ability of the catchment area to neutralize acidity. There is little argument that the pH of poorly buffered lakes in southern Scandinavia and eastern North America has decreased by 1-2 units during the last 30 years: from levels generally above pH 6 to values below pH 5. In consequence, aquatic life at all levels has suffered. These regions receive acid precipitation (pH < 4.5) and are underlain by granitic or similar bedrock with thin and patchy soils.

Changes in land use and management practices, e.g. replacing birch by spruce in forestry, may affect the acid-base status in soil and consequently the lakes and streams in the area. It is very unlikely, however, that these factors could explain the

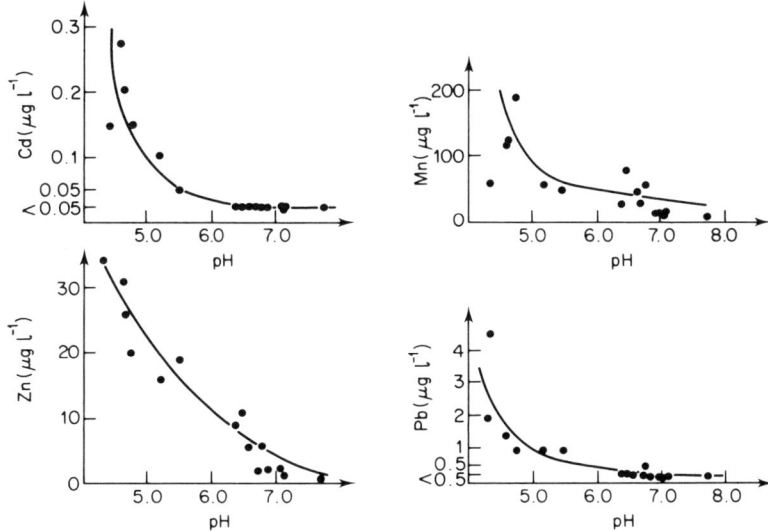

Figure 9.2 Concentrations of metals in 16 lakes with different pH-values on the Swedish west coast. The atmospheric deposition of metals is similar for all lakes. Survey from December 1978 by the Water Research Laboratory, Swedish Environment Protection Board

acidification of freshwater ecosystems in large regions in Norway (Braekke, 1976), Sweden (Ministry of Agriculture, 1978) Canada (Beamish, 1976), and the USA (Likens, 1976). On the contrary the Norwegian SNSF project gives strong evidence for acid deposition as the major source of regional freshwater acidification.

In the SNSF project extensive surveys of the fish populations and chemistry in Norwegian lakes south of 63°N have revealed that the majority of the most acidic lakes have lost their trout populations in recent years. Small lakes at higher altitudes are generally first affected. Similar reports of fish extinction in acidified regions are known from Sweden, Canada, and the USA. No other environmental factor than water acidity seems to explain the gradual regional loss of these fish populations. Indications are that egg and fry mortality are the main cause for failing reproduction, but during large decreases in pH, particularly in spring snowmelt and during heavy autumn rain, acute fish kills have been observed. The tolerance to acid stress in fish depends on many factors, such as content of salts in the water, fish age, size, and genetic background.

Acid lakes are often characterized by higher concentrations of aluminium, manganese, zinc, and cadmium as compared to concentrations of these elements in otherwise similar but less acid lakes. Results from Swedish measurements given in Figure 9.2 can be explained by increased leaching of these metals from the soil. Indications of a significant deterioration in soil chemistry are supported by findings of acid ground water with high metal concentrations in some areas affected by

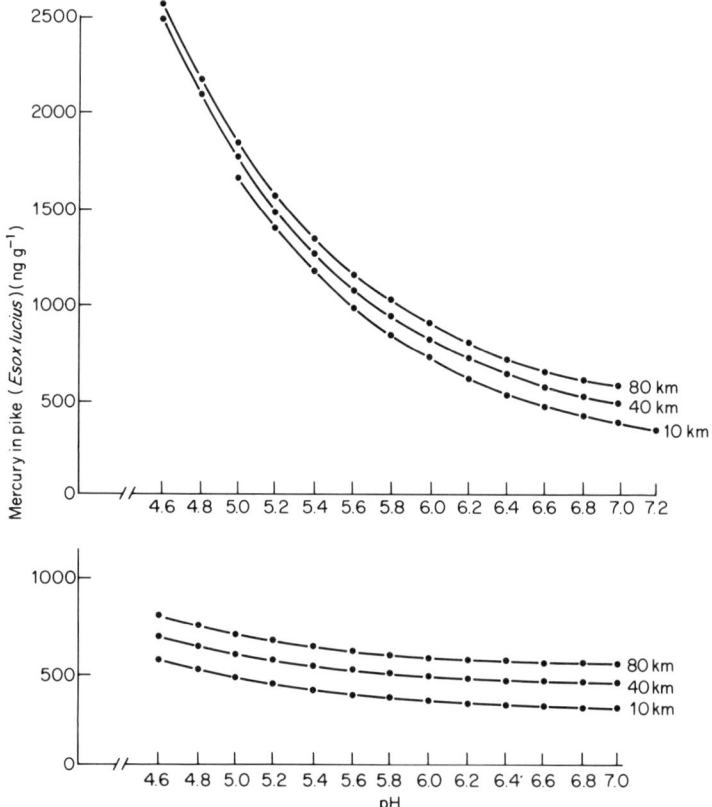

Figure 9.3 Concentrations of mercury in pike (*Esox lucius*) (muscle wet weight) in lakes with different pH-values. The upper curves represent 25 lakes at various distances from a chlor-alkali plant. The curves below represent 31 lakes in a reference area at various distances from the Swedish west coast (Björklund and Norling, 1979)

acid deposition. Aluminium is known to be toxic to fish in the pH range 4.5-5.5 (Schofield, 1977). Mobilization of metals like cadmium and lead as a consequence of acidification must be taken very seriously (Almer et al., 1978).

High concentrations of mercury in fish are correlated with low pH values in water as shown in Figure 9.3. There may be several reasons for this. One important factor is that pH affects the concentrations of individual mercury complexes in water. Such a change in the natural chemical equilibrium may have an impact both on the transfer of mercury between water and air and on the uptake of mercury in fish. The main objectives of the Swedish Environment Protection Board's research programme on mercury are to clarify the effects of acidity on the biogeochemical cycle of mercury and to explain the high concentrations of mercury in fish.

9.3 APPROACHES TO ESTIMATE THE SOCIO-ECONOMIC IMPACTS OF ACID DEPOSITION

The economic and recreational impacts related to the widespread loss of fish stocks may be expressed in monetary terms. In Sweden the area affected by acidification has been estimated to be about 100 000 km^2 comprising 20 000 lakes with a total area of 500 000 ha. The cost due to a total loss of productivity of fish-stocks caught in freshwater and a decreased productivity of fish-stocks caught in the open sea, but breeding in freshwater, has been estimated to be $20-30 million per year (OECD, 1980).

Sport fishing is one of the most popular out door activities in many countries. The acidified lakes and streams in Sweden are of special interest for sport fishing as they are situated in the most populated areas. A total loss of sport fishing has been valued at $50-100 million based on a number of 500 000 sport fishermen in the affected area willing to pay $100-200 per year. The willingness of each fisherman to spend this amount of money was determined in an independent study.

The economic damage caused is important regionally and locally rather than nationally. The money spent on tourism and fishing is of particular importance in certain areas where the local population relies on a combination of farming, forestry, part-time fishing, and tourism for its income. The disappearance of the income from fishing and tourism may jeopardize the livelihood of the local people.

An attempt to evaluate the cost of the destruction of these freshwater ecosystems can be made by calculating the costs for restoring the damaged ecosystems. Liming of affected areas has been proposed as a way to counteract acidification. Based on a test programme launched in 1977, the costs of liming acidified lakes in Sweden have been estimated at $45-75 per ton of limestone (Dickson, 1979). The annual amounts needed to compensate for the acid deposition would be about one million tons of limestone at a total cost of $45-75 million. To compensate for the total acid deposition from the last 30-40 years would cost $1500-3500 million.

Until now anthropogenic acid deposition from sulphur emissions has not been shown to affect adversely the growth of plants except in the vicinity of emission sources. There is, however, a growing concern among scientists for the longer term effects especially on forest productivity. The difficulties in evaluating the effects of acid deposition on forest productivity are related to the fact that it takes 70-100 years for a forest to reach its commercial value. In addition nitrogen oxides contribute to acidification but have at the same time a fertilizing effect. The effect on forestry of deposition of acids has to be kept under review as even a small negative change in productivity has considerable economic consequences.

The economic impacts of the effects of man on the biogeochemical sulphur cycle can also be estimated from a very different starting point.

If we accept that the proved and potential ecological effects of acid deposition are so serious that they cannot be tolerated in the long term we can calculate the costs of controlling anthropogenic sulphur emissions to a 'tolerable' level. Such an approach is justified as the present emissions are in conflict with the declaration of the UN Conference on the Human Environment in Stockholm 1972 which says:

> States have, in accordance with the Charter of the United Nations and the principles of international law, the sovereign right to exploit their own resources pursuant to their own environmental policies, and the responsibility to ensure that activities within their jurisdiction or control do not cause damage to the environment of other States or of areas beyond the limits of national jurisdiction.

What is a 'tolerable' level of sulphur emission in Europe? If we look at the growth rate of the emissions we find that they were fairly constant, around 25 million tons sulphur dioxide per year, during the period 1919-1950. The acidification of aquatic and terrestrial ecosystems seems to be linked to the growing emissions after the second world war. Therefore, the upper limit of 'tolerable' sulphur dioxide emissions in Europe as a whole may be set at 25 million tons. The actual figure is probably lower and will depend on the emissions of nitrogen oxides as they contribute to acidification. It is not necessary, however, to determine the figure with any accuracy for the moment because the control of sulphur emissions has to be a stepwise procedure. The first step must be to stop the increase of the emissions.

There are a number of technical solutions available to reduce sulphur dioxide emissions. The obstacles are mainly of an economic nature.

9.4 METHODS AND COSTS OF SULPHUR DIOXIDE ABATEMENT

The available methods for the control of sulphur dioxide emissions can be classified under give general headings.

(i) Fuel desulphurization processes, in which sulphur is removed from the fuel while the essential nature of the fuel remains unchanged.

(ii) Fuel conversion processes, where the physical state of the fuel is changed, thereby making possible the removal of sulphur.

(iii) Combustion techniques, where an additive is injected which combines with the sulphur.

(iv) Flue gas desulphurization, where sulphur is absorbed from the flue gas produced by combustion.

(v) Fuel substitution, where a high sulphur fuel is replaced by one of lower sulphur content; the replacement of fossil-fired power stations by nuclear energy also falls in this category.

9.4.1 Fuel Desulphurization

For obvious reasons, coal washing is best located close to the source of coal, so that reject material produced in the washing process may be disposed of in the coal mine. In addition centralized coal washeries facilitate the control of any pollution or environmental impact which might arise from this form of fuel desulphurization.

Table 9.1 Typical Cost of Fuel Oil Desulphurization

Desulphurization capacity:	barrels per day tons per year $\times 10^6$	25 000 1.25	50 000 2.4	100 000 5.0
Installed cost ($ $\times 10^6$)		52	80	132
Production Costs ($ year $\times 10^6$)				
Feedstock loss (inc. hydrogen)		4.6	9.1	18.2
Utilities		2.7	5.5	11.0
Catalyst		1.1	2.1	4.2
Maintenance		1.4	2.2	3.6
Labour		15.6	24	39.6
Total		25.4	42.9	76.6
Premium $/ton fuel oil		21.4	18.4	15.6
Additional power cost mils/kW h		5.3	4.5	3.8

Basis: (1) sulphur content of fuel 3.0% sulphur, desulphurization to 1.0% sulphur; (2) hydrogen unit and sulphur recovery system included in capital costs; (3) offsites taken as 20% of process units; (4) capital charge of 30%; (5) all costs in 1978 dollars, instantaneous erection; (6) 8300 operating hours per year.

Furthermore, coal washing is only worthwile if a significant portion of the sulphur in the coal is in the form of pyrites, which is amenable to removal by washing.

Gas cleaning must be accompanied by a means of sulphur recovery. Such installations, for reasons of economy and practicality, should therefore be located at the gas production facilities. In this way, only gas low in sulphur is produced, and this is suitable for either large or small scale combustion applications.

Oil desulphurization is essentially a chemical process, and this is not well known to utility companies. In addition to the desulphurization operation, both a source of hydrogen and a sulphur recovery system are required. For these reasons, oil desulphurization units usually form part of an oil refinery, where the operation of such a unit can be combined with the operation of the refinery as a whole. Such an installation would produce fuel oil to meet specifications for a variety of applications, ranging from large utility installations to smaller scale applications. Typical costs of fuel oil desulphurization are given in Table 9.1.

9.4.2 Fuel Conversion Processes

Coal liquefaction processes will only be justified on practical and economic grounds if built on a large scale where the fuel that is produced can serve a multiplicity of installations. It is probable that fuel produced in this way may be substituted directly for residual fuel oils derived from crude oil. Liquefied coal is thus a possibility for use in both existing and new installations based on fuel oil.

Both coal and oil gasification may be done on a scale small enough to produce fuel for a single installation. Provided that the gasification step is oxygen based, the fuel gas produced could be substituted for coal or oil in existing installations. However, installations of this type are complex, involving in addition to the gasifier, an oxygen plant and sulphur recovery facilities. It remains to be seen how readily such installations will be accepted by the utility industry.

Fluidized bed gasification is suitable for large or small installations, and has the advantage that it is applicable to both new and existing installations. The spent lime produced is regenerable by oxidation to lime and sulphur dioxide, or alternatively the calcium sulphide may be converted to gypsum for eventual disposal. Sulphur dioxide in reasonable quantities can be used in industry.

No costs are presented here as it is felt that these methods are so indeterminate, particularly in the European situation, that any attempt to define costs at this stage would be misleading.

9.4.3 Combustion Techniques

Modified combustion techniques offer considerable promise, and can be applied to almost any fuel. In addition nitrogen oxides can be reduced at low costs by these techniques. They are, however, only suitable for new units, and then only on a reasonably large scale so that the economic benefits of increased size may be obtained.

9.4.4 Flue Gas Desulphurization

The technique of flue gas desulphurization may be applied to almost any boiler or power plant, provided that the physical space necessary for the installation of the desulphurization equipment is available. It is probable, however, that flue gas

Table 9.2 Typical Cost of Flue Gas Desulphurization Processes

Type of Process	Throwaway			Regenerative		
Unit capacity (MW)	100	500	1000	100	500	1000
Capital cost ($ $\times 10^6$)						
New unit	18	52	82	36	104	168
Retrofit	25	73	123	50	146	246
Production Cost (mils/kW h)						
New unit						
Capital charges	6.8	3.9	3.1	13.6	7.8	6.2
Operation	2.0	2.0	2.0	3.4	3.4	3.4
Total	8.8	5.9	5.1	17.0	11.2	9.6
Retrofit						
Capital charges	9.4	5.5	4.6	18.4	11.0	9.2
Operation	2.0	2.0	2.0	3.4	3.4	3.4
Total	11.4	7.5	6.6	21.8	14.4	12.6

Basis: (1) coal fuel, 90% of sulphur removal; (2) 7000 hours/year operation, equivalent to 4500 hours peak load; (3) capital charge 17%, including maintenance; (4) all costs in 1978 dollars.

desulphurization will only be applied to installations of reasonable capacity, in order to obtain the benefits of scale on cost.

The question of waste disposal is of fundamental importance in selecting the type of flue gas desulphurization system to be employed. The throwaway system based on lime or limestone, must have a means of disposing of the waste sludge over the lifetime of the unit. This requirement will probably limit the application of such systems in many urban areas. The typical cost of flue gas desulphurization processes is given in Table 9.2.

9.5 CONCLUDING REMARKS

The first international convention on transboundary air pollution was signed in mid-November, 1979, in Geneva within the framework of the United Nations Economic Commission for Europe (ECE). It is a step forward that the problems of long range transport of air pollution have been recognized by all ECE countries. The member countries have not, however, been able to agree on a programme to reduce the emissions of sulphur dioxide. It is stated in the convention that the parties shall

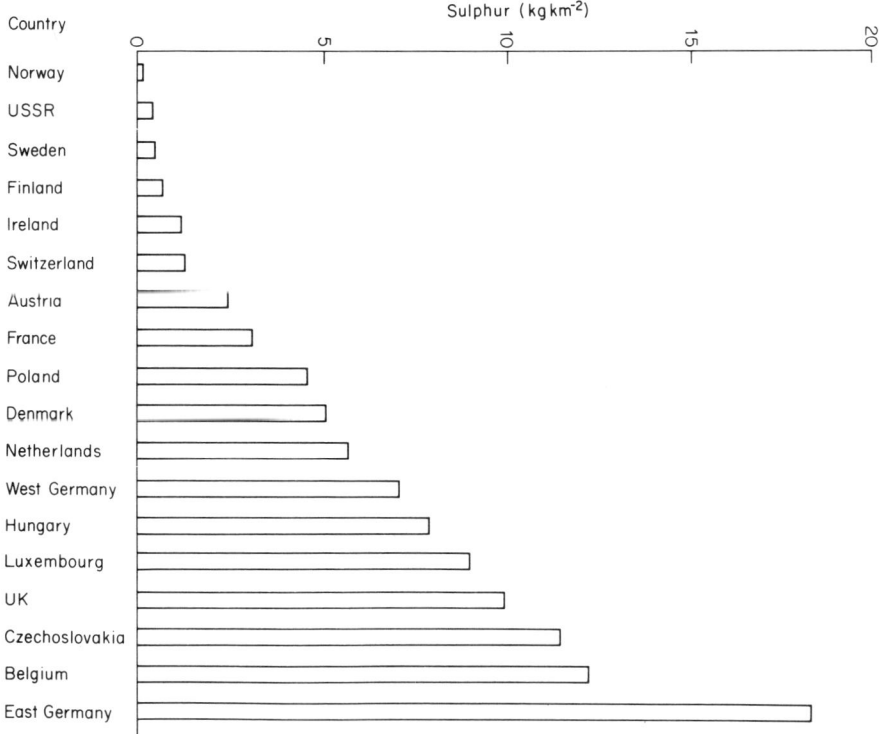

Figure 9.4 Emission density of sulphur dioxide from combustion of fossil fuels in European countries 1978

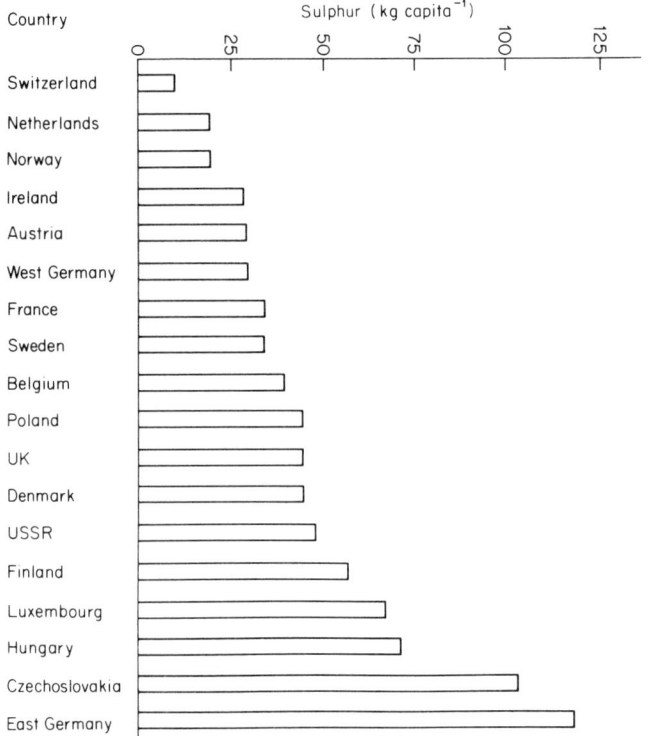

Figure 9.5 Emission of sulphur dioxide per capita from combustion of fossil fuels in European countries 1978

endeavour to limit and, as far as possible, gradually reduce and prevent air pollution including transboundary air pollution. Priority will be given to sulphur oxides.

Great variations exist in emission density and in emission per capita of sulphur dioxide within the European region. Figures 9.4 and 9.5 are based on available emission inventories (Dovland and Saltbones, 1979) for sulphur dioxide from combustion of fossil fuels. Emissions from industrial processes add about ten per cent to these figures.

The costs for individual countries assuming different reduction programmes have been calculated in a recent Swedish study (Ministry of Agriculture, 1979). Two examples of expenditures country by country expressed in dollars per capita are given in Figures 9.6 and 9.7.

Taken together Figures 9.4-9.7 illustrate the great difficulties in formulating a common reduction policy for the European region.

A reduction in sulphur dioxide emissions in Europe from the present level of about 60 million metric tons to 25 million tons per annum by desulphurization of fuels and stack gases would cost at least US $10 billion annually (1978 dollar). It is

Effects of Man on Biogeochemical Cycles: Sulphur 155

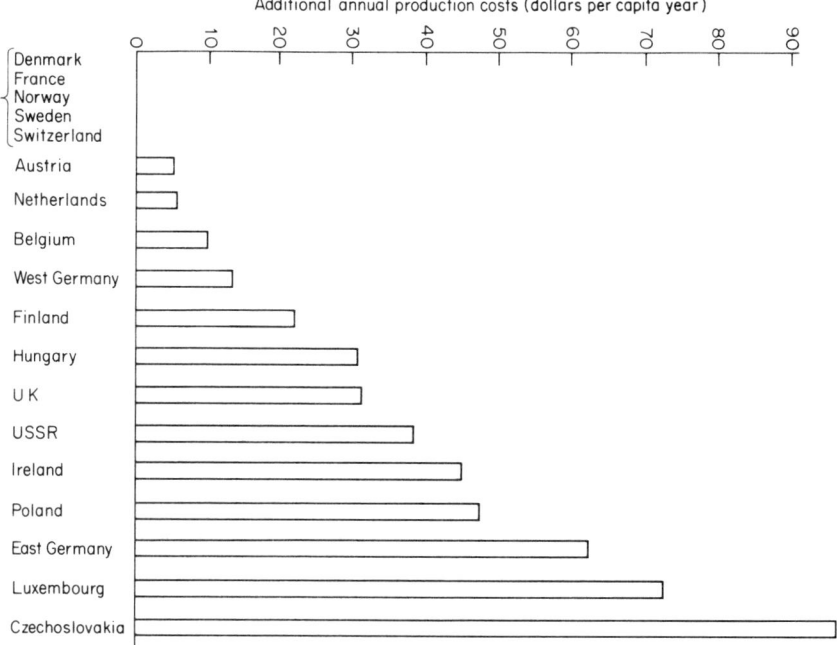

Figure 9.6 Annual costs in European countries to reduce 1990 emissions to 25 kg S per capita by a combination of fuel oil desulphurization and flue gas desulphurization in some coal fired installations

an enormous amount of money. However, if the costs of desulphurization of heavy fuel oil (Table 9.1) are related to the costs of fuel oil, and the costs of fuel gas desulphurization (Table 9.2) to the costs of electricity production the increases are about 10 and 15 per cent, respectively. Additional costs in the range of 10–15 per cent give another perspective of the expenditures involved especially in the light of a 300 per cent increase in oil prices in recent years.

9.6 REFERENCES

Almer et al. (1978) Sulfur pollution and the aquatic ecosystem, in Nriagu, J. O. (ed.) *Sulfur in the Environment, part II, Ecological Impacts,* New York, Wiley.

Beamish, R. J. (1976) Acidification of lakes in Canada by acid precipitation and the resulting effect in fishes, in Seliga, L. S., and L. Dochinger (eds) *Proc. First International Symposium on Acid Precipitation and the Forest Ecosystem,* Columbus, Ohio.

Björklund, I., and Norling, L. (1979) *Effects of Air-born Mercury on Concentrations in Pike and Sediments around a Chlor-alkali Plant,* SNV PM 1090, Swedish Environment Protection Board, Solna.

Braekke, F. H. (ed.) (1976) *Impact of Acid Precipitation on Forest and Freshwater Ecosystems in Norway,* summary report on the research results from phase I of the SNSF project, Oslo.

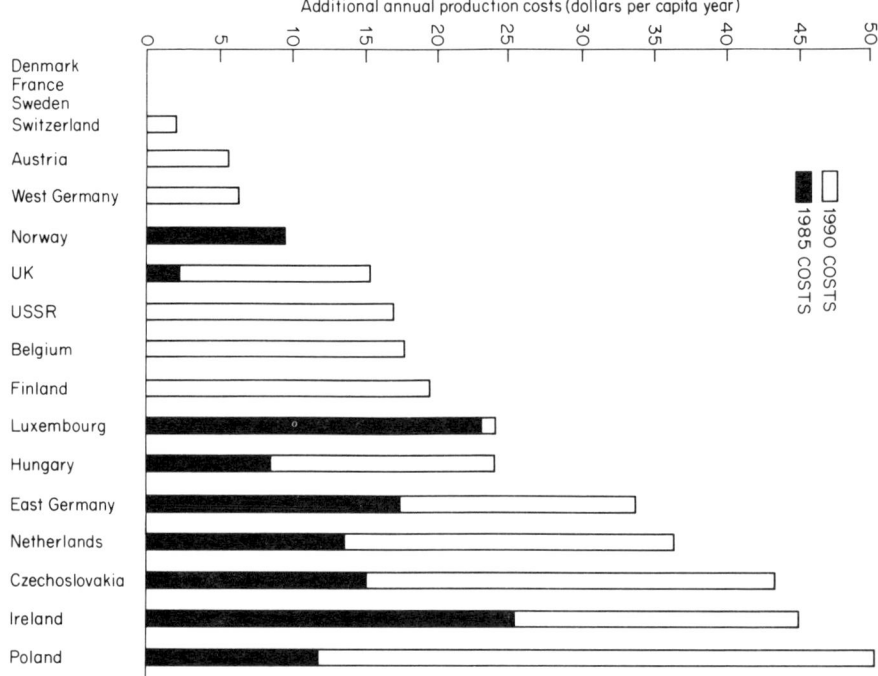

Figure 9.7 Annual costs in European countries to reduce 1985 emissions to equal 1975 levels (filled columns) and to reduce 1990 emissions by 25% from the 1975 levels using a combination of fuel oil desulphurization and flue gas desulphurization in some coal fired installations

Dickson, W. (1979) Experience from small scale liming in Sweden, in *Proc. International Symposium on Sulphur Emissions and the Environment*, London.

Dovland, H., and Saltbones, J. (1979) *Emissions of Sulphur Dioxide in Europe*, EMEP/CCC-Report 2/79, Geneva, Economic Commission for Europe.

Likens, G. E. (1976) Acid Precipitation, *Chem. Eng. News*, **54**(48), 29.

Ministry of Agriculture (1978) *Swedish National Report on Long-Range Transboundary Air Pollution to the Economic Commission for Europe*, Stockholm, Swedish Ministry of Agriculture.

Ministry of Agriculture (1979) *An Investigation into Present and Future Levels of Sulphur Dioxide Emissions in Northern Europe*, Stockholm, Swedish Ministry of Agriculture.

OECD (1977) *The OECD Programme on Long Range Transport of Air Pollutants: Measurements and Findings*, Paris Organization for Economic Co-operation and Development.

OECD (1980) *Managing the Sulphur Problem: Control, Costs and Benefits*, Paris, Organization for Economic Co-operation and Development.

CHAPTER 10

Socio-Economic Impacts of Carbon Dioxide Induced Climatic Changes and the Comparative Chances of Alternative Political Responses – Prevention, Compensation, and Adaptation

K. M. MEYER-ABICH

University of Essen, Essen, Federal Republic of Germany

ABSTRACT

Prevention of climatic change by changing human economic behaviour or compensation for climatically detrimental effects by technological solutions is not necessarily better than adaptation. In fact, there are good reasons to conclude that adaptation is the most rational political option, at the same time requiring least marginal action (i.e. least action specifically for reasons of climatic change). The problems resulting from carbon dioxide all appear at present to be marginal ones which arise, and should be taken care of, for other reasons than simply climatic. With respect to carbon dioxide induced changes, no more action is required than is already necessary for reasons of development policy.

Life in the industrialized countries is to some extent comparable to living in the city of 'Diaspar' in A. C. Clarke's novel 'The City and the Stars'. Diaspar is a triumph of technology, enclosed under a huge dome, a perfect sphere—so to speak —which may neither be left nor entered except for some entropy exchange with the rest of the world. In fact, people in the present-day industrialized societies live remarkably isolated from nature, and this applies especially to weather or climate developments. As a striking example, I may refer to our way of travelling over long distances—e.g. to scientific conferences. We use airplanes which are—for good reasons—air-conditioned; we arrive at an air-conditioned airport, take a few breaths of genuine or even natural 'atmosphere' but then (often enough) enter an air-conditioned taxi which brings us into an air-conditioned hotel or directly into the air-conditioned conference centre. All this may give rise to a certain apprehensiveness particularly when the topic of the conference is the interface between climate and society and when, as once happened to me, moreover in the airplane you read an article stating that agriculture in the future will not necessarily have to be connected with activities outdoors, since what plants need to grow can be supplied to them by nutrient solutions and artificial light as well. Thus even the one basic activity which up to now was conspicuously dependent on climatic conditions could also be transferred, it seems, into the completely artificial environment of the almost-Diaspar in which the industrial societies are generally living and in which, among other things, we are talking about the impact of climate changes on our life.

However, the technological buffer to protect ourselves against the environment is still far from being impermeable or insensitive to outside changes. On the one hand, for many of us, it still makes a considerable difference to work in a room with or without windows, to sense the sunlight and to follow clouds in the sky.

Even if these were only aesthetic experiences they make us feel differently and are important elements of the quality of life. On the other hand, there are also impacts on economic and social activities which perhaps may be most suitably represented by distinguishing different levels of impact between the inorganic sphere, the biosphere, and the sphere of social activity. Such a 'taxonomy of impacts' has been proposed in a former paper (List *et al.*, 1978), the results of which I only very briefly summarize here.

One and the same climatic change basically may be described on five different conceptual levels. These are:

(0)	(1)	(2)	(3)	(4)
Climatic parameters	Environmental potential	Human material activity	Social interaction	Political process

Level (0), of course, refers to the observable atmospheric conditions as described by climatology. At this level, the climatic change in question is determined in terms of temperature, humidity, wind, rainfall, radiation, cloudiness, etc.

At *level (1)*, the environmental potential, a climatic change is described in terms of ecology and geography, considering the impact of the change in climatic parameters on the conditions of life which traditionally are characterized by the four elements soil, water, air, and energy (or light, or 'fire').

At *level (2)* the human species is taken into account as far as our activities in interactions with the environmental potential. Such activities include agriculture, the energy economy, tourism, transportation, construction, manufacturing, etc., and the changes are described primarily in terms of economics and technology. This, however, does not yet include how a given change of climate is expressed in changes of the social interaction among the members of our species. Social interaction is considered at *level (3)*. Effects to be taken into account here are, among others, employment or unemployment, migration, cultural shifts, and social conflcts, so that psychology and sociology may supply a valuable conceptual frame.

Socio-Economic Impacts of Carbon Dioxide Induced Climatic Changes 159

So far nothing has been said about causal relations. A particular change of climate which bears on the weather as well as on the environmental potential, human material activities, and social interaction, in every case can be described in these four different languages or frames of conceptual analysis—no matter whether natural developments or human activities (like burning fossil fuels) have brought about that particular change. The fourfold description may be considered simply as a kind of spectral analysis of one particular development with respect to the field of human knowledge, given only the necessary condition that the phenomenon (also) appears in climatic parameters (while it is not necessary that every change in climatic parameters will correspond to changes in human activities or interactions). Feed-back loops and analyses which consider changes as impacts of other developments (again like burning of fossil fuels), certainly must also be considered an interesting field of scientific enquiry. In the present context, however, causal relations are relevant only insofar as implications with respect to response strategies may be expected, which—as I shall show—is probably not the case with climate.

The political process is *level (4)* of our spectral analysis with respect to climatic changes. On this level climatic changes take the form of policy measures such as legislative action, research and development programme, economic incentives or restrictions, or international policies. Generally the political changes which go together with climatic changes can be described in terms of political science. Even at this level the impact may work both ways, as an impact of climate on political decisions or as an impact of political decisions on climate. Only the effect of *potential* developments is unidirectional in principle, since even the mere possibility of a climate change can elicit a response while possibilities of political developments will not influence climate.

In considering possibilities of CO_2-induced climate changes now, the question is how these changes can be represented in terms of alternative political responses. Of course, such a representation cannot be more reliable than the description of those changes in terms of climatic parameters. Generally it will be much less reliable, because political responses to possibilities of climate change depend on the intermediate representation of those changes on the levels of human activities, material and social. The extreme uncertainty of a political response to the possibility of unknown socio-economic implications of unknown climatic changes, the implications being fairly unknown even if the changes were well-defined with respect to climatic parameters, however, does not imply that it is premature to consider political responses at the given stage of knowledge. The reason is that not only does political action depend on prognostic climatological knowledge, but also the kind of knowledge which can give rise to political action depends on the spectrum of political options. Otherwise every thought in climatology would be politically relevant, which is obviously not the case.

Climatological information, however, will have to be the starting point, even if considerations have to be adjusted back and forth to bring about an adequate assessment. So, what do we know at *level (0)*? In the first place increased CO_2

concentrations in the atmosphere are a matter of intermediate storage since the CO_2 will untimately disappear in the oceans. For the time being, however, it seems to be fairly certain
- that from current and foreseeable developments a rise in global mean temperature by about $2 \pm 0.5\,°C$ must be expected before the middle of the next century as a result of doubling the CO_2 content of the atmosphere by burning fossil fuels and, to a lesser degree, by vegetation destruction and soil deterioration.

This increase in mean temperature would probably not manifest itself significantly at the other levels if it were not secondly connected with
- changes in precipitation patterns. According to Manabe's (AAAS, 1979) latest model, for instance, it is expected that precipitation will increase to the North of 55° latitude and that sizeable changes will also take place in other regions.

Considering the details, however, the state of knowledge is far from satisfactory. Moreover, the given implications are valid (if at all) only as far as anthropogenic effects are considered, while the actual development will be the result of the superposition of anthropogenic effects on natural developments, which nobody is capable of confidently forecasting either. To neglect this distinction gives rise to well-known ambiguities with respect to warming or cooling in the future. The same statement, 'The mean temperature will increase', may refer to the anthropogenic effects alone as well as to the result of the superposition, and while fairly reliably true in the first case is more doubtful in the second case.

According to Flohn (1977) it seems that we also have to expect at *level (0)* a general shift of the climatic zones toward the Poles going together with a melting of the Arctic Sea ice and an enhanced asymmetry between Arctic and Antarctic. Moreover, some (unknown) part of the continental ice in Greenland and in the Antarctic may slowly melt during the centuries of increased temperature in the polar regions. Finally, there is a remote possibility that the West Antarctic ice sheet will break off and cause a surge. This also must be expected independently of CO_2 variations, but possibly with a lower probability.

Even with the state of knowledge with respect to changes in climatic parameters being far from satisfactory, a not completely inconclusive representation at *level (1)* is possible. The result may be, for example, that in the Ruhr area in Germany we are going to have a climate more like the present climate of northern Italy, and— more important—that the habitat for humans in Canada and in northern Russia will be considerably improved. The break-off of the West Antarctic ice sheet may be represented at this level as a rise of the ocean surface by about five metres. In considering the South, it seems that in some regions a decrease in rainfall is to be expected, though so far it cannot be specified in which areas this decrease will occur.

Changes in habitat or in the conditions of life as described in terms of ecology and geography do not only refer to the human species. Our own species, however, is the one whose future we are most immediately concerned with, so that a political response is liable to occur—if at all—most easily if human material activities are affected by climatic change. By this I do not in any way want to suggest that we should not feel responsible for other species. We should feel a responsibility, especially if they are jeopardized by human activities, but not only then. However, given that reasoning with respect to specific developments in this area is very weak so far, it may still be legitimate to conceive in the first place of political responses with respect to the human species and to other species only insofar as they contribute to human welfare. This presupposes a representation of the *level (1)* developments at *level (2)*.

10.1 IMPACTS ON HUMAN MATERIAL ACTIVITIES

The subject at *level (2)* is human activity, not restricted to social interaction as such, but also involving 'material' impacts on the natural environment. We are concerned with a two-ended relationship, human society being on the one end and the material (more or less natural) environment on the other. Since the state of a relationship is dependent on both ends, it follows that the impact of climatic changes on this relationship—i.e. on human activities—depends on the structure of the particular society as well as on the kind of climatic change. As an example: it has been suggested by Burton *et al.* 'that the most vulnerable societies are neither the poorest and least developed, nor the wealthiest and most highly developed, but those societies in the process of rapid transition or modernization, where the traditional social mechanisms for absorbing losses and sharing them among the community has been eroded away and have not been replaced by the accumulated wealth and response capacities of more modern societies' (Burton, 1979). The ability of a society 'to bounce back', as Burton put it, when adversely affected by climatic effects, for example, is called *resilience*. Again, the resilience of a society does not occur in isolation but always refers to specific impacts, as a 'resilience (or vulnerability) with respect to . . . '. Warrick *et al.*, for instance, have argued that even if a livelihood system is insulted from recurrent climatic *fluctuations* as a result of increasingly elaborate technical and social systems, these in turn 'increase the vulnerability to *catastrophe* (my italics) from both natural and social perturbations of rarer frequency' (Warrick *et al.*, 1979).

Understanding resilience and vulnerability as a conditional property of a society with respect to external changes means that these qualities depend on the technology and, it seems to me, on the religious basis of a society, not less than on the particular character of, for example, climatic changes. Different vulnerabilities may result in a twofold disadvantage for the developing countries, since not only is their technological potential less advanced than that of the industrialized countries but also it must be expected that the CO_2-induced changes in climatic parameters for the developing countries have to be represented at *level (2)* as a more severe change in agricultural productivity than in the industrialized world. In fact, the European peoples are not only privileged with respect to wealth but proposed changes in the mean temperature by $\pm 2\,°C$ appear less likely to influence precipitation patterns in a way that will cause serious change in agricultural productivity. Most of the developing countries, on the other hand, are highly vulnerable to even minor variations in water supply, since they are already working at the margin of productivity, water being the main limiting factor of vegetation all over the world except for a few areas, among which only the humid tropics are in the developing world. To some extent the argument also applies to the industrialized world, for instance, since Kazakhstan in the USSR is situated in a semiarid area, and food surplus areas of the USA and Canada are vulnerable to changes in rainfall. On the whole, however, it may be concluded that

- the rich countries not only own the more advanced technologies to cope with changes in agricultural productivity but they are less liable to get into a situation where these means would have to be applied;
- while the poor countries not only lack the technological response capacity they also are the ones which would probably need this capacity.

Finally because agriculture is a relatively minor part of the economic activity of many industrialized countries, while in most of the developing countries the whole economy is still basically dependent on agriculture, it must be concluded that the conflict potential between North and South could be considerably enhanced by CO_2-induced climate changes, because the already existing inequalities in distribution of wealth may tend to be increased.

Apart from agricultural losses—which within a few years may amount to a very considerable proportion of the former productivity if precipitation falls below a minimum—the developing world may also be affected by changes in marine productivity and by losses in (rainfall dependent) hydroelectricity.

Considering the industrialized countries, I certainly do not want to suggest that they will be unaffected. Obviously we feel climatic variability—and as a first order approximation the same would apply to climatic change, or variation—in food prices, tourism, heating expenses, etc. Also the construction business (buildings, roads, etc.), like agriculture, is dependent on seasonal changes. However, in the long run the basic dependency of mankind's cultural and economic development on climatic changes, as has been suggested by Bryson (1978) for the past, in my opinion most probably does not apply to the future of the industrialized world. One cold winter, for instance, may be bad for the construction industry, but many cold winters or any other climatic changes including increases in variability may even be an economic incentive, especially if corresponding information is available early.

So much about *level (2)*. At *level (3)*, the representation of climatic changes in social interaction and international relations has already been assumed to be a rise in the conflict potential between North and South.

At *level (4)*, the remaining question is, what kind of political response(s) would be appropriate to the climatic changes in question?

Appropriateness with respect to a political response generally means that its political 'cost' does not exceed its political benefit. Naturally these costs cannot be expressed in monetary terms but are to be understood as social costs in the broader sense introduced by Kapp, so that alternatives have to be compared in several dimensions. These alternatives are, as Corbett (1979) put it, basically the adaptive, the curative, and the preventive approaches. Adaptation means that nothing is done against the climatic changes in question but that it is our task to fit our activities into a changing climatic pattern. Prevention is to be understood as suspending or restricting the activities which are responsible for undesirable implications. The third approach comprises all possibilities between these two and perhaps should most suitably be called compensation—especially by 'technological fixes'.

Compensation, however, is open-ended with respect to prevention and to adaptation, so that the trichotomy is not satisfying so far. On the one hand, prevention of climatic change may be achieved by originally preventing additional CO_2 generation as well as by preventing CO_2 emission from burning fossil fuels into the atmosphere (alternative disposal of stack gases) as well as by preventing increases of the atmospheric CO_2 content (compensation of additional sources by additional sinks). On the other hand, adaptation to climatic change may range from passive acceptance to global efforts of compensating for the (potential or actual) undesirable impacts of that change on all peoples. To avoid an overlapping between prevention and compensation as well as between compensation and adaptation, in my view it seems reasonable to define

(i) Prevention as prevention of CO_2 generation (for example, by substituting alternative energy forms for fossil fuels);
(ii) compensation as a suspension of undesirable effects of CO_2 generation by global efforts (international activities and budgets);
(iii) adaptation as responses to undesirable effects of climatic change on the national or individual level.

The specific difference between the first two strategies is, therefore, given by CO_2 generation or non-generation, while the two latter strategies differ in the level of response and payment (international versus national or individual). The reason to draw the second distinction with respect to the coordination and payment level is that climate is international and adequate countermeasures against climatic changes must be internationally coordinated, while programmes below the international level can be only piecemeal reactions and may suitably be called adaptation. On both levels, however, anticipatory as well as subsequent reactions are possible.

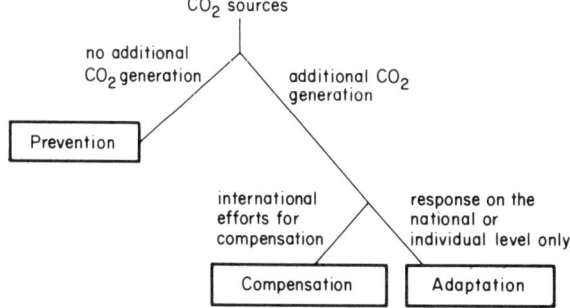

It is important to recognize that prevention, compensation, and adaptation are basically equivalent options, so that it would be a mistake to consider, for instance, prevention as being in principle better than adaptation. What is 'better' only depends on the (social) costs involved (except for taking into account the distinction between 'better for mankind' and 'better for nature', including mankind). The distinction between the three strategies also indicates two basic problems connected with any political response to potential as well as actual climatic changes

(i) The one is that a decision can hardly be identified as the relatively optimal solution without referring to what a future optimum would look like. Since the different parties—especially in international relations—usually do not agree on common goals, the 'least marginal action' (least change in behaviour) will be favoured, which in most cases is expected to be adaptation.

(ii) The other is that the different options generally, or in most cases, may turn out to be options of different parties involved so that for instance, the question is whether country A takes a step for prevention or whether country B takes a step for adaptation. International activities for compensation will also be viewed differently by the different parties involved. Obviously, international bargaining does not necessarily bring about the best solution, which would be the objective of an international authority.

Finally, the three options—prevention, compensation, and adaptation—are connected with quite different time scales. Prevention by its very definition takes place much earlier than compensation, while adaptation again generally may be left to a later time than compensation. Therefore, by way of discounting future costs to the present, the already given bias in favour of adaptation strategies will be even more enhanced with consideration of time.

10.2 PREVENTION OF CO_2 GENERATION

Can we expect that the structural bias in favour of adaptation will be counterbalanced by comparative savings of prevention or compensation strategies in relation to the costs of adaptation—be it only in the interest of the whole, an entity which is not represented by any of the different individual parties? I fear that even a hypothetical international authority charged with making the choice between the different alternatives would not decide in favour of prevention. Two basic reasons are that

(i) there are already strong national incentives to reduce the consumption of fossil fuels which are only slightly enhanced by uncertain assumptions about climatic changes in distant times at distant places. Or, to be more explicit: if the oil crisis has not convinced the American people that oil should be substituted by conservation (capital investment) and by alternative (carbon or non-carbon) fuels and that this is economically reasonable as well as technically feasible, prospects of climatic changes will not convince them either. Conservation and solar energy should be given priority over nuclear energy insofar as nuclear energy cannot be used to substitute for oil until forms of storage or of transforming nuclear energy into fuel are developed. As far as coal is concerned, CO_2-emission per energy unit is 50 per cent higher than from oil, but this will not prevent the substitution of oil by coal wherever possible. Thus in my judgment energy development within the next 40-50 years will by itself not bring about a four-fold increase in the CO_2 content of the atmosphere, but again by itself cannot be prevented from bringing about a two-fold increase;

(ii) few changes are to the disadvantage of everybody. There are likely to be quite a few countries, including some in the Third World, which are going to receive net benefits from climatic change. Therefore, the political conflict may as a first order effect turn out to be a distributional conflict, even within the Third World, with some countries developing migration pressures with respect to others.

In any case, it is most improbable that the industrialized world will slow down economic activity by energy restrictions for reasons beyond those which already exist and that other countries will stop burning forests to prevent CO_2-induced climatic changes. After all, events like the Sahelian drought or a Peruvian El Niño irregularity should be considered bad enough to elicit action so that the irregular becoming regular probably will not bring about changes in human behaviour either. Different reasons or causes for similar results do not seem to change this situation.

My conclusion, however, is only that prevention should not be expected at any considerable price. To the extent that CO_2-induced climatic changes can be prevented by way of a joint production of benefits, by piggybacking CO_2-oriented measures on to measures which are accepted for others reasons, the costs of which do not rise significantly by bringing about that additional benefit, even preventive steps may reasonably be expected. The joint production of benefits from reducing oil consumption (and thereby perhaps imports) and CO_2 emissions at the same time is a good example, for certain countries, of a piggy-backing strategy.

10.3 COMPENSATION BY INTERNATIONAL EFFORT

If CO_2 generation which would double the CO_2 content of the atmosphere within the next 40-50 years cannot be prevented, the next question is whether something can be done so that at least the undesirable implications will be prevented. This is a field where considerable amounts of technological fantasy and imagination are called for.

First I may refer to Marchetti's famous 'Gigamixer'. The idea is to put the stack gases from electricity generation as well as from heat generation centres (by burning fuels with pure oxygen or by filtering out CO_2 and other components) into a current which is at the ocean surface at Gibraltar and then disappears in the deep sea— being supposed not to show up again within the next five hundred years, after which overloading the atmosphere with CO_2 will not be a problem anymore, because by then mankind will have run out of fossil fuels. A similar idea is to use the stack gases as a fluid for tertiary recovery in exhausted oil fields (Marchetti 1979).

Secondly, one may think of replacing rainfall by irrigation, even if this were extremely expensive in terms of money as well as energy.

A third idea is to redesign biological species or their geographical distribution so that agricultural production under changed climatic conditions will be similar to today's production from today's species.

Others have proposed planting 10^{12} trees or 'moving a metre of topsoil from Iowa to where Iowa's climate will then be' (L. Lave at AAAS, 1979) on a global scale. Finally, the most appealing idea to the modern mind—appalling, however, when accepted as a general principle—may be to venture into 'global climate management'. In fact, having agreed on so much resource management already, be it water, energy, or the environment in general, there seems to be no conclusive reason why our domination of nature and our treatment of nature as a 'resource' should not be extended to climate. On the other hand, we are beginning to realize that technological solutions of problems are generally tied up with social commitments and that there are good arguments in favour of the recommendation to consider these social commitments not less but even much more carefully than the technology itself (Meyer-Abich, 1979).

Apart from these more philosophical considerations, the chances of compensation measures depend on the costs involved and on agreement among the different parties as to who will be charged which share of the total costs. Considering the price of technological fixes, Marchetti has estimated that electricity production would become about 30 per cent more expensive if the stack gases were transferred to his gigamixer. If one takes into account that new technologies tend to be π times more expensive than originally calculated, this may be a large amount of money. Global climate management—basically cloud and rain distribution management—will not necessarily be a less expensive solution. These costs, however, are not high or low by themselves but are high or low only with respect to the benefits in question.

The benefit of preventing climatic changes by technological compensation may be given as the opportunity costs minus the benefits of a climatic change (if there are any). The opportunity costs are to be understood as the additional benefits which would have occurred if the climate had not changed, or as losses brought about by climatic change. Within the Climate Impact Assessment Program (of the US Department of Transportation), d'Arge (1974) has ventured to calculate such opportunity costs for some parts of the global economy, on the basis of a scenario with a 1 °C decrease in mean annual temperature. Pretending any accuracy and reliability at all for such partial or further extended calculations has been strongly criticized by Margolis (1978), and I agree with him. This criticism even applies to calculating opportunity costs of climatic changes with respect to present activities with given objectives in a given climate. It applies, therefore, even more

- *to future* activities the goals of which are unknown, so that nobody knows whether climatic changes can be held responsible for missing political goals to this or that extent or whether they contribute to achieving them. Or, if goals were made explicit, again it would be an open question whether they will be missed on account of climatic changes. A country, for instance, might claim to have been prevented by climatic changes from becoming the wealthiest country in the world. It may well be that this claim, 'iffy' as it is, could hardly be refuted; to stating deviations with respect to a reference case which is ill-defined in itself, since the climate is also changing by natural developments. The benefit of compensating for anthropogenic effects on climate obviously is ill-defined when natural fluctuations of the same order of magnitude may be expected which nobody can predict or exclude so far.

Finally it must be pointed out that even if the opportunity costs and benefits of climatic changes were known or at least conceptually defined, the political problem of charging different parties with costs of the technological fixes in relation to their particular responsibilities as well as to their costs and benefits would be practically insurmountable. Climate cannot be nationalized. It is essentially an international concern, so that any climate management or compensation strategy involves economic externalities—positive or negative—with respect to national borders. Investments to compensate for climatically harmful activities, therefore, will be almost without returns generally if not endeavoured on the basis of international cooperation, excluding 'free rider' policies as much as possible. The implication is that something like the 'polluter pays principle' from environmental policy also should be applied in climatic matters. It is at this point and so far only at this point that the question of different national liabilities with respect to climatic changes arises. Since not even the benefits of technological compensation strategies can be determined, however, the question of cost distribution according to differential national liabilities may be considered irrelevant. Also it may be argued that costs should be distributed according to the expected benefits instead of the shares in pollution.

10.4 ADAPTATION TO CLIMATIC CHANGE

As Glantz (1979) has pointed out, climatic change by CO_2 production would be another of those low grade, but continually increasing, insults to the environment for which a pluralistic society '. . . has not yet found an effective policy-making process'. If the problem—as it seems—cannot be taken care of by prevention or technological fixes (compensation), the chances of adaptation depend on developing policies with respect to 'impending crises'. As Mann (1979) put it: 'What we need is not a massive decision but a gradual learning process'. This process may lead to

- migration into those regions which are favoured by the climatic changes in question. Though it is very hard to live as an immigrant, this is the traditional solution to such problems in the history of mankind;

- vocational re-education and industrialization. This is going to happen anyway and will also allow for an increase in population density or make up for decreases of agricultural productivity.

Obviously, the adjustment of economies adapted to the present climate to a different climate and the migration of hundreds of millions of people again imply considerable costs. These costs are highly dependent on early information about the developments to be expected so that forecasting climate—allowing for active rather than passive adaptation—can save enormous amounts of money, and they will occur only within decades. In the context of present development debates, however, it seems that with respect to the next 40–50 years

- neither are we confronted with a new problem, since the overburdening of productivity capacity by high population densities is happening already and has been a problem for a long time;
- nor are any monetary claims involved which would change the present situation as far as development policies are concerned.

Compared to the already existing problems in development policy, the possibility of CO_2-induced climate change, therefore, seems to be a 'marginal' problem in the sense of not being qualitatively different, while quantitatively not significantly increasing the already given tensions. For example, the recommendation to increase food reserves (requiring higher production in the industrialized world), reasonable as it is with respect to possible climatic changes and food shortages, is eminently reasonable with respect to the present situation as well. The same applies to Elise Boulding's (1979) recommendation 'to draw on skills that are now hidden from policy makers', reviving 'traditional knowledge stocks of peasant and nomadic communities, of ethnic groups in industrialized societies, or minority-status groups in all societies, including particularly women and children'.

Of course, nobody can exclude that crop yields in some developing countries will drop by, say, 50 per cent within five years for reasons of climatic change. A fifty per cent decrease in productivity corresponds to a hundred per cent increase in population density, and this—at a rate of two per cent per annum—would be reached only in thirty-five years, so that the assumed climatic change comes out to be something like 'seven times worse' than that population increase. But again, even if we knew such a drop in crop yield were to be expected at some future time, we could not do better than do what should be done already for reasons independent of CO_2.

So far the gist of my argument is that even an international authority with definite political objectives—except for piggy-backing strategies—would not decide in favour of

- prevention because the already given incentives are only marginally enforced by the climatological argument, and because further prevention has not been shown to be better than non-prevention on a global scale;
- compensation because the benefit of those expensive programmes is ill-defined, not to mention the distributional problems, while the costs of adaptation are

distant, are again only poorly defined, and in any case are marginally charged on account which is grossly imbalanced anyway.

A comparative cost evaluation of the three strategies, therefore, cannot be conclusive. The situation would be different if for reasons of climatic change catastrophic developments of the same kind and order of magnitude as those which have to be expected now, or undesirable developments of a new quality were to emerge beyond 40-50 years. As Flohn (1977, 1979) and others argue, this may very well be true, even if to some degree uncertain with present knowledge. At the same time, however, it is not a political issue requiring additional decisions now.

Adaptation, therefore, seems to be the most rational political option for the time being. It also requires the least marginal action (i.e. least action specifically for reasons of climatic change and not also or mainly justified for other reasons). No problems have been identified so far with respect to CO_2—or otherwise induced climate change—which change the political situation and which should not be emphasized and taken care of for better and more urgent reasons as well. Politically, the CO_2 problem is like chalk on a white wall—or rather like some additional darkness in the night. To blame only the marginal darkness for the gloom is politically misleading or even deceptive. This result, however, should be taken only as a first order representation of possible CO_2-induced climatic changes at the political level. Much more consideration will have to be given to interactions among the different levels to achieve an adequate assessment. The next step may be to look for climatological representations of political conflicts, taking into account, for instance, that the industrialized world

(i) may be blamed for food shortages in the developing countries because of CO_2 changes, even if there is no sound climatological justification. There is so much talk about CO_2 or other man-induced climatic change that minor climatic irregularities already give rise to the question whether 'this is it' (S. Schneider, personal communication);

(ii) is generally experiencing a growing concern about being responsible for social and technological commitments of mankind in the very distant future, so that the concern about climate may be only one element or a symbol of a more deeply rooted uneasiness.

In this sense, the climatic concern beyond itself may become an important focus for social and political concern about not adequately doing what should already be done for reasons which have nothing to do with CO_2.

10.5 ACKNOWLEDGEMENTS

The views presented in this paper resulted from much inspiration as well as criticism received at the SCOPE/CCC workshop on Climate/Society Interface, arranged by F. K. Hare in Toronto, December 10-14, 1978, and at the AAAS/DOE workshop on Environmental and Societal consequences of a possible CO_2-induced climate change, Annapolis, April 2-6, 1979. The paper itself has been carefully reviewed by

Jesse Ausubel. Comments and valuable criticism have also been given by Hermann Flohn and Ulrich Hampicke. I am grateful for everything that I have learnt and may be blamed for what I have not learnt.

10.6 REFERENCES

American Association for the Advancement of Science (1979) *Workshop on Environmental and Societal Consequences of a Possible CO_2-induced Climate Change,* Annapolis, April 2–6.

d'Arge, R. (1974) *Contribution to the Climate Impact Assessment Program (CIAP),* Washington, D.C., Department of Transportation DOT-TST-75-50.

Boulding, E. (1979) Reflections on the mandate of panel IV, Social and institutional responses to a global CO_2-induced climate change, in *Workshop on Environmental and Societal Consequences of a Possible CO_2-induced Climate Change,* AAAS.

Bryson, R. A. (1978) Cultural, economic and climatic records, in Pittock, A. B. *et al.,* (eds), *Climatic Change and Variability,* 316–327.

Burton, I. (1979) Social and behavioural responses to climatic change and variability: The role of perception and information, in *Workshop on Environmental and Societal Consequences of CO_2-induced Climate Change,* AAAS.

Corbett, J. G. (1979) Developing response strategies for climate change, in *Workshop on Environmental and Societal Consequences of CO_2-induced Climate Change,* AAAS.

Flohn, H. (1977) Stehen wir vor einer Klima-Katastrophe?, *Umschau,* **77**, 561–569.

Flohn, H. (1979) Eiszeit oder Warmzeit? *Naturwissenschaften,* **66**, 325–330.

Glantz, M. H. (1979) A political view of CO_2, in *Workshop on Environmental and Societal Consequences of a CO_2-induced Climate Change,* AAAS.

Hare, F. K. (1979) Report on the SCOPE Workshop on Climate/Society Interface, Toronto.

Kapp, K. W. (1950) *The Social Costs of Private Enterprise,* Cambridge, Massachusetts.

List, R., Meyer-Abich, K. M. and Williams, W. (1978) The conceptual spectrum of climatic impacts, summarized in Hare (1979).

Mann, D. (1979) Lecture at the Annapolis workshop, AAAS.

Marchetti, C. (1979) *Constructive Solutions to the CO_2 Problem,* Laxenburg, International Institute for Applied Systems Analysis.

Margolis, H. (1978) *Estimating Social Impacts of Climate Change – What might be done versus what is likely to be done,* SCOPE.

Meyer-Abich, K. M. (1979a) *Energieeinsparung als neue Energiequelle* wirtschaftspolitische Möglichkeiten und alternative Techologien, München, Hanser Verlag, (in German).

Meyer-Abich, K. M. (1979b) Towards a practical philosophy of nature, *Environmental Ethics,* October 1979.

The National Research Council (1978) *International Perspectives on the Study of Climate and Society,* Washington, D.C., The National Research Council.

Scientific Committee on Problems of the Environment (1975) *Workshop on Climate/Society Interface,* Toronto, December 10–14, cf. Hare (1979).

Warrick et al. (1979) The effect of climate fluctuations on human populations, in *Workshop on Environmental and Societal Consequences of CO_2-induced Climate Change,* AAAS.

Index

Acetylene 33, 34
Acid–base status 146
Acid deposition 146–48
 freshwater ecosystems 149
 socio-economic impacts 149–50
Acid lakes 147
Acid rain 104
Acidification 121
Adaptation 167–69
Aerosol injection 13
Aeschynomena indica 35
Agricultural productivity 161–62, 166, 168
Agung, Mount 13, 14
Air conditioning 157
Akcagylian marine transgression 11
Albedo perturbations 5, 10
Algae, nitrogen-fixing blue-green 114
Alnus glutinosa 35
Aluminium 121, 139, 140, 147, 148
Ammonia 27, 29, 38, 44, 82–84, 105
Ammonium 27, 29–31, 33, 41
Ammonium oxidase 34
Anabaena azollae 34
Anaerobic metabolism 36
Anaerobic microsites 37
Aphanizomenon 116
Aphanizomenon flos-aquae 115
Aphanizomenon gracile 115
Aragonite 137
Aral, Lake 11
Atmospheric circulation model 11
Atmospheric cycles
 chemical coupling of 81–91
 trace gases in 81–91
Azolla 34, 35
Azospirillum 34, 37
Azotobacter paspali 34

Baltic Sea 73
Biogeochemical cycles
 interactions in terrestrial ecosystems 93–112
 marine ecosystems 125–42
 quantitative evaluations 97
 socio-economic impacts on 143
Biomass accumulation 95
Black Sea 11, 12, 72

Cadmium 105, 139, 140, 147
Calcareous shells 137
Calcite 137
Calcium 74
Calcium carbonate cycle, marine ecosystems 136–38
Carbon 17, 44, 87, 100, 108, 117, 120, 121
 organic 74, 126–31
Carbon-14, concentration and generation 16
Carbon compounds 81
Carbon cycle 71, 83, 98, 100–3
 effects on cycles of other elements 120–21
 marine ecosystems 126–31
 phosphorus cycle effects on 116–17
 sulphur cycle effects on 119–20
Carbon dioxide 19, 64, 74, 97, 100, 101, 105, 116, 117, 119, 159, 160, 162–65, 168, 169
 and climate 17–20
 atmospheric level 17
 man-made 19, 21
 socio-economic impacts 157–70
Carbon flux 102
Carbon monoxide 82, 83, 87, 89, 90
Carbon:nitrogen ratio 30, 100
Carbon:nitrogen:phosphorus ratio 101
Carbon:phosphorus ratio 100
Carbonate compensation depth (CCD) 18
Carbonyl sulphide 55

Caspian Sea 11, 12
Catastrophe vulnerability 161
$(CH_3)_2S$ 82
CH_4 82, 83, 87, 89
Chemical coupling
 atmospheric cycles 81–91
 reactions in 83–91
Chemosynthesis 64
Chlorides 68, 69, 105
Chlorophytes 116
Citrobacter freundii 35
Climate management 166
Climatic changes 5, 158, 162–69
 extraterrestrial 20, 21
 terrestrial 21
Climatic effects 161
Climatic factors 3–24
Climatic fluctuations 161
Climatic parameters 158–60
Climatological information 159
Clostridium perfringens 31
Coal washeries 150
Coccolithophores 136
Combustion of fossil fuels 102
Combustion techniques 152
Compensation 163, 165
Computer model studies 9
Continental configuration 9
Copper 105
COS 82
Cretaceous shoreline 6
Crustal movement dynamics 19
Cyanobacterium 35
Cyanophyceae 114

Datisca cannabina 35
Decomposition 99
Deglaciation 5
Denitrification 31–34, 36, 37, 121
Desiccation 5
Deuterium/tritium isotope 16
Developing countries 161, 168, 169
Diaspar City 157
Dissolved inorganic carbon 116, 119
Drainage waters 106, 107
Dust particles 13

Earth's rotation rate 10–11
Ecosystems
 functioning of 38
 size effects 95
Eh effects 129

Environmental potential 158
Environmental protection 158
Enzyme systems 33
Eocene transgression 8
Epilimnion 118
Epilithiphyton 115
Erosion 17, 99, 100
Eustatic sea-level changes 5
Eutrophication 114
Experimental Lakes Area 115

Fertilizers 27
Flooding 5
Flue gas desulphurization 152
Fluidized bed gasification 152
Flux determination 94
Foraminifera 136
Forest growth, simulation models 102
Forest products, harvesting 102
Fossil fuels 17, 102
Frankia 35
Free energy changes 33
Freshwater ecosystems 113–23, 149
Fuel conversion 151
Fuel desulphurization 150–51
Fuel gas purification 75, 76

Galactic model 4
GAMETAG experiment 54
Gas cleaning 151
Geotectonic model 5–11
Gigamixer 165, 166
Glaciation 5
 Paleozoic 10
 Precambrian 4, 11
Glutamate dehydrogenase (GDH) 27
Glutamine synthetase/glutamate
 synthase (GS/GOGAT) 27
Gunnera 35

Harvesting of forest products 102
H_2CO_3 104
Hippophaë rhamnoides 35
HO_2 84, 86, 87, 88
H_2O_2 86
HO_x 87, 88
H_2SO_4 83
Hubbard Brook Ecosystem Study 106
Hubbard Brook Experimental Forest
 95, 99, 100, 104, 107, 108
Human material activities 158, 161–64
Hydrocarbons 82

Index

Hydrogen 102
Hydrogen sulphide 52, 64, 69, 71, 72, 74, 76, 82
Hydrologic cycle 98–100
Hydrologic fluxes 100
Hydroxamic acids 33
Hydroxylamine 31
Hypolimnion 118, 120

Invertebrates 94
Iron 99, 121
Iron sulphide 118, 120

Kinetic models 126
Kinetic theory 95
Klebsiella pneumoniae 37
Krakatoa 13

Lake level oscillations 16
Land mass movements 10
LD-50 experiment 114
Leaching 39, 99
Lead 99, 105
Learning process 167
Long Island 98
Lysocline 18, 19

Magnesium 102
Manganese 147
Marine ecosystems
 biogeochemical cycles 125–42
 calcium carbonate cycle 136–38
 carbon cycle 126–31
 nitrogen cycle 132–35
 phosphorus cycle 131–32
 silica cycle 135–36
 trace elements 138
Mass balance 94, 97, 125–26, 135, 136
Mauder minimum 15
Mercury 148
Meteorological parameters 56
Michigan, Lake 118
Microorganisms 97
Milankovitch curve 12–13
Mineralization 30, 44, 74, 97, 99
Myrica gale 35

NH_2 83, 84
NH_3 27, 29, 38, 44, 82–84, 105
Nickel 139
Nitrate 29, 31, 33, 38, 41, 56, 102, 106–8, 121

Nitric acid 83, 120
Nitrification 31–34, 36, 37, 44, 97, 102
Nitrite 32
Nitrobacter 32
Nitrogen 105, 108, 118, 121
 assimilation 27
 circulating in vegetation-microorganism-soil system 40
 complexity 25
 dissolved 135
 distribution in biosphere 26
 inorganic 132
 man-made 27
 valence states 25
Nitrogen budgets 41
Nitrogen compounds 27, 44, 81
Nitrogen cycle 25–49, 83, 105–6
 ecosystem 38–41
 effects on cycles of other elements 120–21
 global 27, 41
 in soil 33
 interactions 41–44
 marine ecosystems 132–35
 microniche concept 35–38
 phosphorus cycle effect on 114–16
 subcycles 82
Nitrogen fixation 27, 34–35, 38, 44, 114, 135
Nitrogen flux 134
Nitrogen immobilization 39
Nitrogen losses by leaching 39
Nitrogen metabolism 27–35
Nitrogen oxides 27, 56–58, 82, 83, 87–90, 105, 106
Nitrogen:phosphorus ratio 114, 115
Nitrogenase 34
Nitrogenous substances 32
Nitrosomonas 31, 32, 36
Nitrous oxide reductase 34
Nitroxyl 32
Nutrient deficiencies 114
Nutrient flux 98–99
Nutrient input/output and fate 108
Nutrient relationships in terrestrial ecosystem 96
Nutrient release 100
Nutrient supply 104

O_3 84, 89
OH 87–91

OH-radical 83, 84, 86, 87, 90
OH recycling 87
Oil desulphurization 151
Oligocene regression 8
Oocystic sp. 116
Opal production 136
Orbital periodicities 12–13
Organic matter 136, 137
Oxygen 74, 89
Oxygen cycle 64

Paleozoic glaciations 10
Paspalum notatum 34
pH effects 44, 102, 103, 116, 129, 146, 148
Phosphate 102, 121, 139
Phosphorus 44, 99, 105, 121
Phosphorus cycle 106–8
  effects on carbon cycle 116–17
  effects on nitrogen cycle 114–16
  effects on silicon cycle 118–19
  effects on sulphur cycle 117–18
  freshwater ecosystems 113–14
  marine ecosystems 131–32
Photic zone 127, 134–36
Photo-assimilation 64
Photosynthesis 64, 98, 116, 126, 129
Phytoplankton 115, 127, 135, 139
Planetary alignments 15
Plate accretion and subduction 19
Plate motions 9
Plate tectonics 5
Political process 158, 159
Political responses 159, 162
Pollution standards 105
Potassium 102
Precambrian glaciations 4, 11
Precipitation 56, 103, 104, 106, 161
Pycnocline 19

Quaternary 11, 13

Redfield ratio 101
Regressions 5–10
Resilience 161
Respiration 129
Rhizobium 34, 35, 37

Salmo gairdnerii 115
Savanna ecosystems 35
Scenedesmus 115, 116
Sea floor spreading 6

Sea-level fluctuations 6, 19
Sedimentation 7, 8, 73
Signal-to-noise ratios 95
Silica cycle, marine ecosystems 135–36
Silicon 139, 140
Silicon cycle, phosphorus cycle effects on 118–19
SNSF project 147
SO_4 106
SO_4^{2-} 106
Social interaction 158
Socio-economic impacts
  acid deposition 149–50
  carbon dioxide 157–70
  on biogeochemical cycles 143
  sulphur 145–56
Sodium 102
Soil animals 30, 32
Soil moisture effects 36
Solar activity 15, 16
Solar flux 16
Solar parameter 14
Source–receptor relationship 57
Spatial distribution of continents 9–10
Sport fishing 149
Stratification 19
Stratified waters 20
Stream water 107, 108
Subsidence 7, 8, 11
Sulphate 56–58, 64, 68, 69, 71, 72, 102, 107, 108, 119, 145–46
Sulphides 64
Sulphur 73, 105
  biogenic 71
  biologic 70
  contamination sources 75
  in industrialized regions 56–57
  organic 72
  production 75
  runoff 68
  socio-economic impacts 145–56
Sulphur budget 73
Sulphur compounds 44, 52–54, 59, 63, 64, 71, 73, 75, 81
Sulphur cycle 64, 102–5
  anthropogenic fluxes 65–67
  atmosphere 51–60, 63, 69–71
  effects on carbon cycle 119–20
  global biogeochemical 61–78
  in ocean 71–74
  interactions with other cycles 58, 104

Index

man-made contributions 74–76
outlook for the future 58–59
phosphorus cycle effects on 117–18
present state of problem and objectives 62–65
quantitative evaluation of other cycles from 74
Sulphur dioxide 51, 54, 58, 82, 87, 90
  control methods 150–52
  distribution 55
  emission control 152
  emission control costs 154–55
  emission density 153
  emission per capita 154
  fuel gas purification of 75
Sulphur emission 150
Sulphur fluxes 54, 55, 61–66, 70, 72, 74, 75
Sulphur oxides 56–57, 69, 145
  fuel gas purification from 76
Sulphuric acid 64, 75, 97
Sulphurous aerosols 105
Sunspot cycles 16

Tambora, Mount 13
Tectonic pulses 11–12
Tectonism 19
Terrestrial ecosystems 94–97, 106
Thermodynamic modelling 126
Thermodynamic parameters 95
Thermodynamic stability 132
Third World 165
Throughfall 99
Tidal forces 15
Trace elements, marine ecosystems 138
Trace gases in atmospheric cycles 81–91
Transgressions 5–10
Transpiration 98

Van, Lake 16
Vegetation 102, 104
Volcanic activity 17, 19, 69
Volcanic dust 13–15

Water cycle 41, 98–100
Water effect 36, 99
Watersheds 95, 97, 106
Weathering 17, 64, 95, 104
Weertman model 13
Wind directions 56

Zinc 105, 139, 140, 147